Economic and Medicinal Plant Research

Volume 2

Economic and Medicinal Plant Research

Volume 2

Edited by

H. WAGNER
Institut für Pharmazeutische Biologie
der Universität München
München, West Germany

HIROSHI HIKINO
Pharmaceutical Institute
Tohoku University
Sendai, Japan

NORMAN R. FARNSWORTH
Program for Collaborative Research in
the Pharmaceutical Sciences
College of Pharmacy, Health Sciences Center
University of Illinois at Chicago
Chicago, Illinois, U.S.A.

1988

ACADEMIC PRESS
Harcourt Brace Jovanovich Publishers

London San Diego New York
Boston Sydney Tokyo Toronto

ACADEMIC PRESS LIMITED
24/28 Oval Road,
London NW1

United States Edition published by
ACADEMIC PRESS INC.
San Diego, CA 92101

British Library Cataloguing in Publication Data

Economic and medical plant research
Vol. 2.
1. Plants, useful —— Research
I. Wagner, H. (Hildebert) II. Hikino, Hiroshi III. Farnsworth, Norman R.
581.6′1′072 QK98.4

ISBN 0–12–730063–5

Phototypeset by Photo·graphics, Honiton, Devon, England
Printed in Great Britain

Contents

3
Natural Products for Liver Diseases

Hiroshi Hikino and Yoshinobu Kiso

4
Potential Fertility-regulating Agents from Plants

Audrey S. Bingel and Harry H. S. Fong

Contributors

Numbers in parentheses indicate the pages on which the authors' contributions begin.

AUDREY S. BINGEL (73), Program for Collaborative Research in the Pharmaceutical Sciences, College of Pharmacy, Health Sciences Center, University of Illinois at Chicago, Chicago, Illinois 60612, U.S.A.

GÁBOR BLASKÓ (119), Program for Collaborative Research in the Pharmaceutical Sciences, College of Pharmacy, Health Sciences Center, University of Illinois at Chicago, Chicago, Illinois 60612, U.S.A.

GEOFFREY A. CORDELL (119), Program for Collaborative Research in the Pharmaceutical Sciences, College of Pharmacy, Health Sciences Center, University of Illinois at Chicago, Chicago, Illinois 60612, U.S.A.

NOEL J. DE SOUZA (1), Hoechst India Limited, Mulund, Bombay 400080, India.

HARRY H. S. FONG (73), Program for Collaborative Research in the Pharmaceutical Sciences, College of Pharmacy, Health Sciences Center, University of Illinois at Chicago, Chicago, Illinois 60612, U.S.A.

HIROSHI HIKINO (39), Pharmaceutical Institute, Tohoku University, Sendai 980, Japan

YOSHINOBU KISO (39), Pharmaceutical Institute, Tohoku University, Sendai 980, Japan

VIRBALA SHAH (1), Hoechst India Limited, Mulund, Bombay 400080, India

H. WAGNER (17), Institute of Pharmaceutical Biology, University of Munich, Munich, West Germany

Preface

The varied character of natural plant products, and indeed their very existence, pose fundamental questions to scientists. Many books have been published concerning the chemical aspects of these products; however, it is exceptional to find discussed within a single volume most aspects of particular genera or of particular pharmacological classes of natural substances, all having economic potential.

Thus, the intent of this book series is to identify areas of research in natural plant products that are of immediate or projected importance from a practical point of view and to review these areas in a concise and critical manner.

We feel that these topics will be of great interest to graduate students, research workers, and others interested in the discovery of natural products and in their further utilization as drugs, pharmacological tools, models for synthetic efforts, or other economic purposes. We hope decision makers in industry, government agencies, philanthropic foundations, and elsewhere will benefit from these timely reviews and consider these and related projects as worthwhile endeavours for further research.

1

Forskolin—An Adenylate Cyclase Activating Drug from an Indian Herb

NOEL J. DE SOUZA
VIRBALA SHAH
Hoechst India Limited
Mulund, Bombay 400080
India

I. INTRODUCTION

Forskolin is a labdane diterpenoid isolated from the Indian herb *Coleus forskohlii* (Willd.) Briq. Its remarkable chemical and biological properties were brought to the attention of the international scientific community through a series of publications during 1977–1978 by the group at the Indian Hoechst Centre for Basic Research and by their collaborators at Hoechst AG, West Germany (de Souza, 1977; Bhat *et al.*, 1977; Lindner *et al.*, 1978).

ECONOMIC AND MEDICINAL PLANT RESEARCH VOLUME 2
ISBN 0-12-730063-5

In the early years of the discovery of this natural product, not only was the association of its cardiovascular activity profile with its structure considered to be of high interest, but also its biochemical property to activate adenylate cyclase in a receptor-independent manner was seen to be a dramatic new finding with immense potential for use as a research tool in cyclic cAMP-related studies (de Souza, 1977; Metzger and Lindner, 1981a,b). Forskolin sky-rocketed into international prominence following the demonstration by Seamon and Daly that the uniqueness of forskolin's adenylate cyclase activation lay in its action at the catalytic subunit of the enzyme, an ability not possessed till then by any known substrate of the enzyme (Seamon *et al.*, 1981; Seamon and Daly, 1981a). Global interest in forskolin is reflected in the rapid growth in publications on the molecule since 1981, which now total up to about 600. Vistas were opened for utilizing forskolin as a drug for diseases in addition to those of the cardiovascular domain arising from its discovery—diseases attributed to cyclic AMP deficiency or to aberrations of cyclic AMP metabolism. During the past few years, conclusive evidence has been provided of the ability of forskolin to lower intraocular pressure, to display cardiotonic activity, and to relieve bronchoconstriction (Rupp *et al.*, 1986).

In kindling a resurgent interest in the search for novel medicinal agents from plants, *C. forskohlii* is now immortalized through its inclusion among such historical medicinal plants as *Atropa belladonna*, *Cinchona ledgeriana*, *Rauvolfia serpentina*, *Digitalis lanata*, and *Dioscorea deltoidea*. Different reviews describe the biochemical and biological properties of this unique diterpenoid constituent (Rupp *et al.*, 1986; Seamon and Daly, 1981b; de Souza *et al.*, 1983; Seamon and Daly, 1983; Seamon, 1984; Daly, 1984; de Souza, 1986; Seamon and Daly, 1986). This review, in updating the current status and forward thrust of recent studies, will attempt to focus on the input that would be needed to bring out one or more products of forskolin onto the market, thereby making it a reality that *C. forskohlii* not only remains a plant of medicinal interest, but also becomes a plant of high economic importance.

II. HISTORICAL PERSPECTIVE, THE PLANT SOURCE, AND THE CHEMICAL ENTITY

In the quest for novel drugs from Indian plants, the research group at Hoechst India Limited, Bombay, India was led via its screening programme to the isolation, of forskolin from the roots of the herb *C. forskohlii* (de Souza, 1977; Bhat *et al.*, 1977). About 15,000 species of vascular plants are reported to be growing in India, of which over 2000 species are described to be medicinal in the traditional systems of medicine such as Ayurveda, Siddha, and Unani. The figure of 2000 is further supplemented by the ethnomedical plants used by the 300 or more Indian tribes.

Selection of plants for the screening programme was based on a combination of medical, chemical, and botanical criteria (de Souza *et al.*, 1983). *Coleus forskohlii* was one such plant that was collected because of its phylogenetic relationship to *C. amboinicus* Lour, a species equated to the traditional Ayurvedic and Unani drug, Pashanbhed. It has since become known from literature reports and oral communications that the species has several ethnomedicinal uses not only in India but also in Nepal, Zanzibar, and Pemba (Williams, 1949). A root decoction is used as a tonic by the Kota tribe of Trichigadi village near Kotagiri, in the Nilgiri Hills in South India (Abraham, 1981). The roots are also used in the treatment of worms in Gujarat (Bhambhadai, 1951). A root paste is claimed to allay burning in festering boils. When mixed with mustard oil, the ground root is applied to eczema and skin infections. The plant is also used for veterinary purposes. Roots cooked in water and mixed with feed are given to sick cattle having a loss of appetite for food and water. Furthermore, the green leaves cooked in water are given to cattle to increase lactation (H. Latwal, personal communication).

Coleus forskohlii belongs to the family Labiatae (Lamiaceae), a family of mints and lavenders known to have predominantly diterpene constituents. The genus was conceived by de Loureiro from the Greek word COLEOS meaning sheath, referring to the fused filaments of the flower that form a staminal sheath around the style (de Loureiro, 1790). The epithet *forskohlii* was given by Willdenow to commemorate the Finnish botanist Forskal, who travelled extensively in Egypt and Arabia on plant surveys and died in 1790 of malaria in Arabia. *Coleus forskohlii* (Willd.) Briq., *Plectranthus barbatus* Andr., and *P. forskohlii* Ait. are synonymous names of the plant, for which the legitimate name is *Coleus barbatus* Benth (Bruce, 1935). *Coleus forskohlii* grows on the sun-exposed, dry hill slopes between an altitude of 300 and 1800 m. It is distributed over the subtropical, warm temperate climatic zone on the mountains of India, Nepal, Sri Lanka, Africa (Mukherjee, 1940) Burma and Thailand. It is a herbaceous species with perennial root stock. The stems grow erect up to 45–60 cm in length and become decumbent when grown longer. The inflorescence and flowers are typical of the family Labiatae. The roots are fasciculated, succulent, and tortuous or radially spread (Figs. 1 and 2). The annual availability of fresh roots from their natural habitat in India is assessed to be about 1000 t. The plant has been under cultivation in India in the states of Gujarat, Maharashtra, and Karnataka for use in pickles and condiments.

FIG. 1 *Coleus forskohlii* whole plant with tortuous tuberous roots.

Forskolin is 7β-acetoxy-8,13-epoxy-1α,6β,9α-trihydroxylabd-14-en-11-one (Fig. 3). The chemical and spectral data in support of its structural and stereochemical features have been adequately documented in the literature (Bhat *et al.*, 1977). More recent data obtained from difference Nuclear Overhauser Enhancement spectra has confirmed the α-orientation of the 7-proton following irradiation at the 8-CH_3, and the β-orientation of the 1-proton following irradiation at the 10β-CH_3. A two-dimensional correlated spectroscopy spectrum of forskolin vividly reconfirmed the connectivities of the coupled protons (H. W. Fehlhaber, personal communication). The absolute configuration of forskolin was confirmed by X-ray analysis of forskolin and of its 1-benzyloxy-7-deacetyl-7-bromoisobutyryl derivative (Paulus, 1980a,b). The identity of the structure of

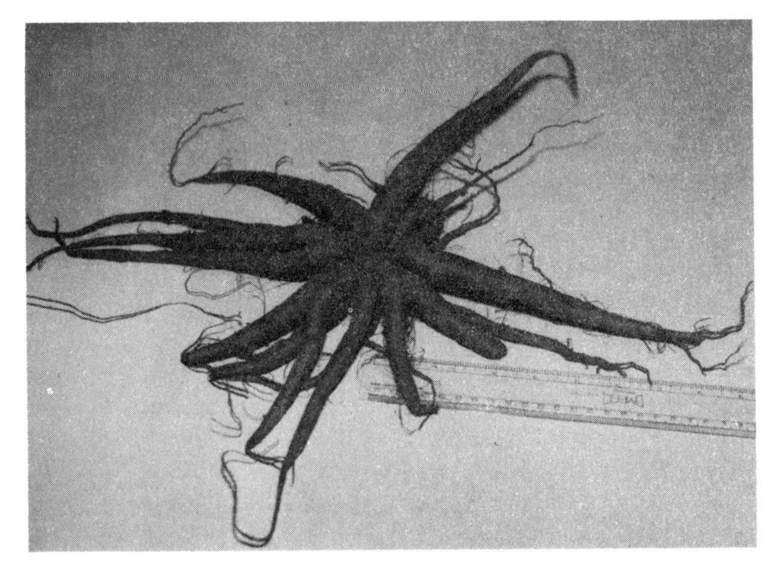

FIG. 2 *C. forskohlii* radially-spread tuberous roots.

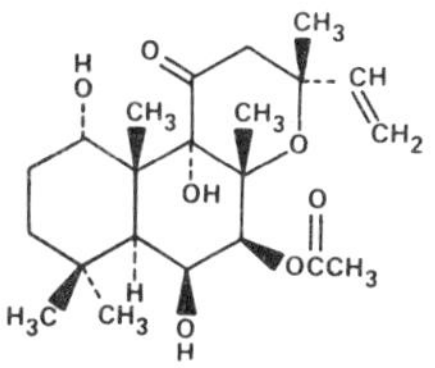

FIG. 3 Structure of forskolin.

coleonol (Tandon *et al.*, 1977) with forskolin has recently been established (Ramakumar *et al.*, 1985; Saksena *et al.*, 1985).

Several unusual structural features characterize this labdane-type diterpene metabolite of the plant *C. forskohlii* growing in India. Unlike the previously reported abietane-type metabolites, such as barbatusin and cyclobutatusin, isolated from the same species growing in Brazil (Wang *et al.*, 1973, 1974), and the coleons isolated from other East African *Coleus* and related species (Arihara *et al.*, 1975), the A-ring hydroxy substituent in forskolin is at C-1 and the C-7 substituent has the β-configuration.

Thin-layer, gas–liquid, and high-performance liquid chromatographic methods were developed for the assay of forskolin (Inamdar *et al.*, 1980, 1984). Via application of these methods to crude plant extracts, the occurrence of forskolin was found to be unique in the plant *C. forskohlii* growing in different parts of India. Forskolin could not be detected in six other *Coleus* species and in six taxonomically-related *Plectranthus* species at levels down to 1×10^{-4}% of the dry weight of the plant material (Shah *et al.*, 1980).

A search for forskolin and adenylate cyclase activators from medicinal plants available in Japan was also made (Kanatani *et al.*, 1985). One-hundred samples were screened, of which three were *Coleus* species and the others belonged to species of the genera *Orthosiphon* and *Plectranthus* of the subfamily Ocimoideae to which *Coleus* belongs. Although eight samples of Japanese plants showed marked stimulation of adenylate cyclase, none was as potent as the methanol extract of *C. forskohlii* roots from India and, more pointedly, none was found to contain forskolin as a constituent. The Indian herb *C. forskohlii* thus remains, even today, the sole known natural source of the diterpenoid, forskolin.

III. FORSKOLIN, ADENYLATE CYCLASE, AND CYCLIC AMP-DEPENDENT PHYSIOLOGICAL EFFECTS

The unique activation of cyclic AMP-generating systems by forskolin has been the subject of different publications and reviews (Metzger and Lindner, 1981a,b, 1982; Seamon and Daly, 1981b, 1986; Seamon, 1984; Daly, 1984).

A. INTERACTION WITH THE CATALYTIC SUBUNIT

Forskolin continues to provide insights into the regulation of the multi-component enzyme complex, adenylate cyclase. The hormone-sensitive adenylate cyclase system is composed of at least five separate subunits: stimulatory (R_s) and inhibitory (R_i) hormone receptors, two separate guanine nucleotide-binding regulatory components, N_s and N_i, acting as couplers between the stimulatory and inhibitory hormone receptors, respectively, and the catalytic subunit C, which actually catalyses the formation of cyclic AMP and ATP (Fig. 4).

Many hormones are capable of regulating this enzyme in either a stimulatory (H_s) manner or an inhibitory (H_i) manner, thus modulating the levels of intracellular cyclic AMP and eliciting the appropriate physiological responses. Forskolin has been shown to

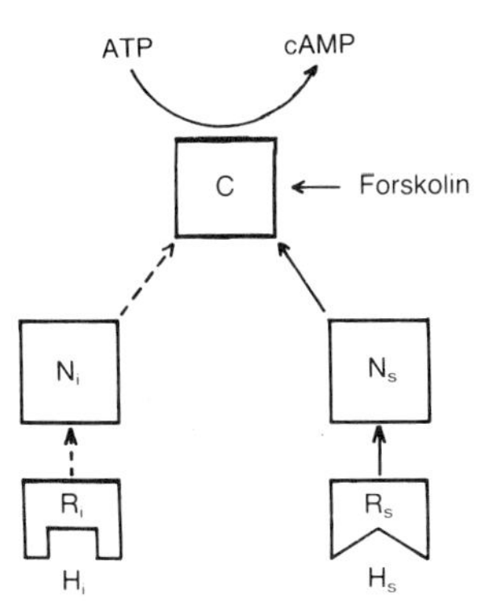

FIG. 4 The multicomponent adenylate cyclase complex.

activate almost all hormone-sensitive adenylate cyclases in intact cells, tissues, and even solubilized preparations of adenylate cyclase, with EC_{50} values usually in the 5–15 μM range. The unique feature of this activation is that the site of action for forskolin is the catalytic subunit of the enzyme or a closely associated protein (Seamon and Daly, 1981a). From this unique interaction, several consequences accrue that have a bearing on cyclic AMP-dependent physiological effects. These are adequately documented in the literature (Seamon, 1984; Seamon and Daly, 1986) and are alluded to only briefly in this paper. It also became possible to bind the catalytic subunit C of solubilized adenylate cyclase from rat brain or heart to a 7-*O*-hemisuccinoyl-7-deacetylforskolin–Sepharose affinity column (Pfeuffer and Metzger, 1982; Pfeuffer *et al.*, 1983). The catalytic subunit was then eluted from the column with forskolin. Purification of the catalytic subunit, its eventual characterization, and a reconstitution of the adenylate cyclase system are thus achievable goals (Metzger, 1986).

B. MODULATION OF HORMONE ACTION

Stimulations of adenylate cyclase systems can be inhibited by various hormones that function to inhibit adenylate cyclase via N_i-proteins. Thus, activation by forskolin is not able to overcome the inhibitory influences of N_i input, as with α_2-agonists, for example, by A_1-adenosine agonists, by somatostatin, and by opiates (Seamon, 1984; Daly, 1984).

On the other hand, low concentrations of forskolin ($< 1\ \mu M$), which alone elicit small increases in intracellular cyclic AMP, markedly potentiate hormonal responses. In a similar manner, low concentrations of hormones can potentiate the forskolin-elicited increases in cyclic AMP. Forskolin and hormones are tremendously synergistic, resulting in both a shift in apparent EC_{50} values and in the magnitude of maximal response attained by the hormone of forskolin alone. Thus, although forskolin does not by itself require an N_s-protein for its direct activation, it can obviously interact in a potentiative manner with hormonal input, which does require the N_s-protein. Through binding studies conducted with [12-^{3}H]forskolin and the identification of two classes of binding sites, one of high affinity and the other of low affinity, these observed synergistic effects between forskolin and hormones in activating adenylate cyclase have been explained by a cooperative binding of forskolin and activated N_S-protein at the catalytic subunit (Seamon, 1985).

These insights provided by forskolin into the regulation of the adenylate cyclase enzyme, and the subsequent tremendous surge in reported research studies that described the potential use of the molecule, not only as an invaluable research tool for understanding cyclic AMP-dependent physiological processes but also as a potential therapeutic agent for diseases in addition to those of cardiac insufficiency and hypertension indicated by early studies, namely diseases such as glaucoma, thrombosis, asthma, and metastatic conditions (Seamon, 1984), underscored the intrinsic merit of forskolin to be developed as a drug.

IV. PHARMACOLOGICAL PROPERTIES, PHYSIOLOGICAL EFFECTS, AND THERAPEUTIC POTENTIAL

Different reviews describe the pharmacological profile displayed by forskolin (Lindner *et al.*, 1978; Dohadwalla, 1986; de Souza *et al.*, 1983; de Souza, 1986; Seamon and Daly, 1986), the physiological effects it elicits in different systems, including cultured cells, tissue cells, and isolated organ systems (Seamon, 1984; Seamon and Daly, 1986), and the clinical potential it has for the treatment of congestive heart failure, asthma, and glaucoma (Rupp *et al.*, 1986). The reader is directed to the cited literature for details of the studies that have been conducted. The discussion in this section will attempt to summarize the progress made in such studies that have aimed at putting a therapeutic product of forskolin onto the market.

A. HAEMODYNAMIC AND CARDIAC METABOLIC EFFECTS OF FORSKOLIN

Detailed pharmacological studies established that forskolin lowered normal or elevated blood pressure in different animal species through a vasodilatory effect and that it had a positive inotropic action on the heart muscle (Lindner *et al.*, 1978; de Souza *et al.*, 1983; Seamon and Daly, 1986). The antihypertensive effects of forskolin were attributed to a decrease in peripheral resistance by action of the drug on arteriolar smooth muscle. The positive inotropic effects of forskolin in the cardiac muscle were shown to be initiated by stimulation of the myocardium cell-membrane adenylate cyclase, followed by a rise in cyclic AMP, an activation of membrane-bound protein kinases, a lowering of the membrane-bound Na^+, K^+-ATPase activity, an activation of the slow calcium gate, and an activation of cytoplasmic protein kinases (Metzger and Lindner, 1981a,b, 1982). Forskolin was thus considered to be the prototype of a new class of compounds that exhibit inotropy by this type of stimulation, and was anticipated to be useful as a cardiotonic drug for the treatment of congestive heart failure.

In different clinical studies (Linderer and Biamino, 1986) conducted on two groups of patients, one having congestive cardiomyopathy with a decreased ejection fraction of about 35% at baseline, and the other having coronary artery disease (CAD) with chronic stable angina, the anticipated improvement of cardiac function and increase of myocardial contractility were convincingly substantiated. Illustrative of these studies is the one carried out on four patients with CAD, wherein the heart rate was kept constant at 100/min by atrial pacing and forskolin was administered as an infusion in a dose of 1.25–3 μg/kg body weight for 15 min (Fig. 5). Mean dP/dt increased from 1775 ± 250 to 2150 ± 208 (mmHg/sec), and mean cardiac index (CI) from 3.1 ± 0.4 to 3.5 ± 0.4 ($\ell/min/m^2$). Left ventricular and diastolic pressure (LVEDP) decreased slightly, and mean arterial blood pressure (MAP) remained nearly constant. Additional studies were also conducted in India (Lele *et al.*, 1986).

The following summarized conclusions were drawn on the basis of the different studies:

1. Forskolin reduces preload and afterload of the heart due to its vasodilating action.

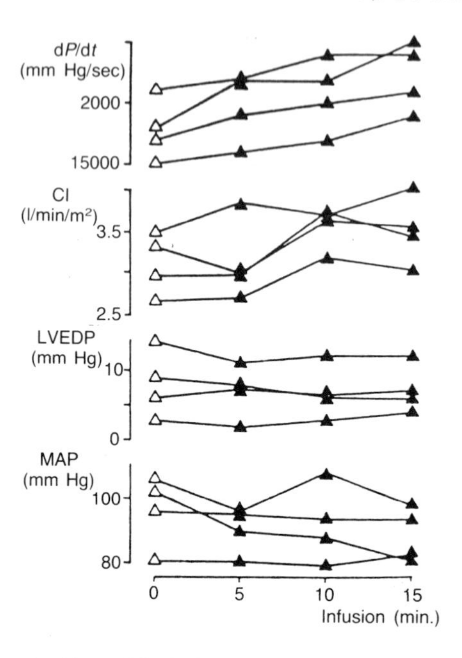

FIG. 5 Haemodynamic effects of forskolin in patients with coronary artery disease. Atrial pacing 100/min. For abbreviations, see text.

2. Forskolin augments myocardial contractility due to its positive inotropic action; it does so without affecting myocardial oxygen consumption.

3. Used in an appropriate dosage, forskolin seems to be a promising drug for treatment of patients with congestive heart failure.

B. BRONCHODILATORY EFFECTS OF FORSKOLIN

Forskolin was studied as a bronchodilator for its potential use in the treatment of asthma. Forskolin relaxed contracted airways *in vitro* and prevented bronchospasm *in vivo* (Burka, 1983a,b; Chang *et al.*, 1984). In addition, forskolin reduced Schultz–Dale responses of both trachea and parenchyma, and protected sensitized guinea-pigs during antigen challenge. Forskolin was shown to inhibit immunological histamine release from chopped sensitized lung. Arachidonic acid metabolism to leukotrienes in the airways could be attenuated by forskolin, indicating a role for it in enhanced mucociliary clearance. Thus, it was demonstrated that forskolin inhibited both mediator release and smooth-muscle contraction in the airways.

Forskolin was also shown to be useful in reducing inflammatory reactions that may contribute to asthma. Forskolin increased cyclic AMP levels in neutrophils and macrophages, and was associated with a decrease in the superoxide burst and prostaglandin synthesis (Burka, 1986).

In an initial study in man, forskolin was administered by a metered-dose nebulizer in amounts of 1 mg and 5 mg to six extrinsic asthmatics with metacholine-induced bronchoconstriction. The forskolin expiratory volume in 1 sec (FEV_1), and airways resistance (R_{aw}) were measured. After forskolin inhalation, FEV_1 rose and R_{aw} decreased in all patients (Fig. 6). No difference was observed between the 1-mg and 5-mg doses. Tolerance was good and no adverse side-effects were experienced. In another double-blind-placebo controlled study, the protective effect of forskolin against acetylcholine-induced bronchoconstriction was investigated in 12 healthy subjects (Kaik, 1986). Forskolin (2 mg and 10 mg) was almost as effective as fenoterol (0.4 mg), but the duration of action of forskolin was shorter compared to fenoterol. An advantage of forskolin over conventional bronchodilators is that it has little or no systemic effects. In summary, forskolin by itself or in combination with longer-acting bronchodilators can be expected to be effective in patients suffering from chronic obstructive lung disease.

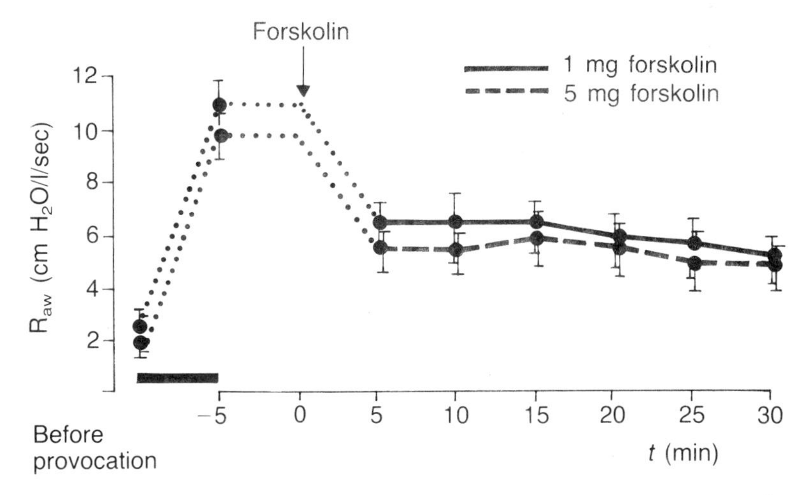

FIG. 6 Effect of forskolin inhalation on R_{aw} (airways resistance) after metacholine provocation.

C. INTRAOCULAR PRESSURE-LOWERING PROPERTIES OF FORSKOLIN

The effect of forskolin on aqueous humour dynamics and intraocular pressure (IOP) was first described by Caprioli and Sears (1983). They demonstrated that topical application of forskolin lowered the IOP in rabbits, monkeys, and healthy human volunteers. In man, 50 $\mu\ell$ of a topical suspension of 1% forskolin produced a 70 ± 16% decrease in outflow pressure in 1 hr, the effect reaching a peak at 2 hr, and remaining significant for at least 5 hr. The reduction of IOP was associated with a reduction in aqueous inflow and no change in outflow facility, indicating a potential for forskolin as a therapeutic agent in the treatment of glaucoma.

The IOP-lowering effects of forskolin were subsequently substantiated in healthy volunteers in the Hoechst laboratories at Frankfurt, at various eye clinics in Germany and Japan (Witte, 1986; Seto *et al.*, 1986), and in patients with open-angle glaucoma in India (Pinto-Pereira, 1986). The drug was well tolerated, with transient effects of hyperaemia, mild sensations of burning, and increased lacrimation. The results of current studies indicate that forskolin has a number of potential advantages in glaucoma therapy: (a) it does not induce miosis; (b) unlike β-blockers, it increases intraocular blood flow; (c) it has no systemic effects; (d) its effect can be additive with other drugs for glaucoma, and possibly potentiated by the use of sympathomimetics (Witte, 1986).

V. PHARMACOKINETICS AND BIOTRANSFORMATION OF FORSKOLIN

The pharmacokinetics and biotransformation of forskolin were studied with [15-^{14}C]forskolin in dogs and rats (M. Volz, H. M. Kellner, and H. G. Eckert, personal communication, 1987). The absorption was complete after oral dosing. In the dog, blood levels reached a maximum 1 hr after administration and decreased very rapidly in the initial phase ($t_{1/2}$ = 2 hr), followed by a significantly slower terminal phase ($t_{1/2}$ = 4 days). In the rat, blood levels reached a maximum between 8 and 32 hr after administration and decreased very slowly, the half-life for the monophasic process being ~13.5 hr. The long-lasting blood-level concentrations indicated a binding to erythrocytes. The preferred route of excretion was via the urine in dogs, and the faeces in rats. Excretion was rapid and nearly complete within 3 or 4 days. The compound was distributed principally in the liver and the kidney, in which organs the concentrations were consistently higher than those in the blood.

Forskolin was metabolized to a considerable extent (M. Volz and H. W. Fehlhaber, personal communication, 1987). Numerous metabolites were isolated from dog and rat urine. Only very small amounts of unchanged drug were found. In dog urine, the biotransformation products of forskolin were mainly present in conjugated form. After cleavage by a mixture of glucuronidase and arylsulphatase (glusulase), the conjugates formed three metabolites, of which the main one was identified as 3β-hydroxyforskolin, yielding 20–25% of the administered dose. The other two metabolites were identified as 3β-hydroxy-7-deacetylforskolin and 3β-hydroxy-7-deacetyl-6-acetylforskolin, in yields of 7–10% and 2–3% respectively, of the administered dose.

VI. ANALOGUES OF FORSKOLIN

Analogues of forskolin have become available through different sources. Some were identified as constituents of *C. forskohlii* (Bhat *et al.*, 1977), over 150 analogues were prepared by semisynthesis (de Souza, 1986; Laurenza *et al.*, 1987), and others were obtained through microbial transformations (Ganguli, 1986; Nadkarni *et al.*, 1986; Khandelwal *et al.*, 1987a). Most of these analogues were tested in different pharmacological and biochemical models, structure–activity relations were derived, and a schematic representation was proposed of the forskolin–receptor interaction between 1α- and 9α-OH groups and the vinyl group of the α-face of the forskolin ring system and the active site of the enzyme (de Souza, 1986; Bhat *et al.*, 1983; Seamon *et al.*, 1983). This model of interaction continues to be used for the design of newer forskolin analogues. Analogues of forskolin more soluble in water than forskolin were synthesized and shown to activate adenylate cyclase, as well as to display pharmacological properties, with potencies and efficacies comparable and even superior to forskolin (Laurenza *et al.*, 1987; Khandelwal *et al.*, 1987b). These water-soluble derivatives of forskolin may be useful for increasing cyclic AMP in broken-cell preparations or in intact-cell preparations, in which the presence of organic solvents, which are necessary to solubilize forskolin, are detrimental (Laurenza *et al.*, 1987). They are also potential second-generation forskolin-derived drugs, which, by virtue of their water-solubility, would meet the objection to the use of the poorly-water-soluble forskolin in intravenous and other galenical preparations requiring an aqueous base.

VII. CONCLUDING REMARKS

Forskolin is clearly the prototype for a novel class of drug with a high potential for therapeutic application in different diseases. In diseases such as congestive cardiomyopathy and bronchial asthma in particular, in which the repeated use of β-adrenergic agonist drugs leads to desensitization or "down regulation" of the receptors and a loss of drug efficacy, it appears attractive to use a drug like forskolin, which acts at sites distal to the β-adrenergic receptor. Additional advantages are likely to accrue from a combination of forskolin with β-adrenergic agonists, cyclic AMP-phosphodiesterase inhibitors, and hormonal agents on the basis of its property to act synergistically with such agents. Development of appropriate galenical formulations and drug delivery systems are likely to provide a product with increased efficacy, longer duration of action, and better patient compliance. The different preparations of forskolin under clinical investigation have of necessity to go successfully through the mandatory steps of the clinical phases I to III before they get the clearance to be introduced in the market place. The challenge remains to introduce the optimal therapeutic product using forskolin or one of its newer analogues, singly or in combination, in a conventional or novel drug delivery form.

On the biochemical front, however, forskolin and its analogues are already in the market place, distributed by companies such as Calbiochem, U.S.A. In view of its potential as a tool for the study of both the biochemistry and regulation of adenylate cyclase and the role of cyclic AMP in physiological functions, forskolin was made available in 1980 to the international scientific community. In 1983, 7-*O*-hemisuccinyl-7-deacetylforskolin was introduced. It can be attached to sepharose to provide an affinity support for the purification of adenylate cyclase, thus enabling the accomplishment of a goal which has eluded researchers for several years. Newly introduced in the market are two analogues, one an inactive naturally-occurring forskolin analogue, 1,9-dideoxyforskolin, which could, serve as a "negative control", and the other a forskolin derivative, about as potent an adenylate cyclase activator as forskolin, with considerably enhanced solubility in water.

The basis for the medicinal and economic importance of the Indian herb *Coleus forskohlii* has clearly been established.

ACKNOWLEDGEMENTS

We wish to thank Dr. R. H. Rupp for his support of this work. We also thank Mrs. A. Nogueira for typing the manuscript.

REFERENCES

Abraham, Z. (1981). *In* "Glimpses of Indian Ethnobotany" (S. K. Jain, ed.), p. 315. Oxford & IBM Publishing Co., Bombay.

Arihara, S., Ruedi, P., and Eugster, C. H. (1975). *Helv. Chim. Acta* **58**, 343.

Bambhadai G. K. (1951). "Vanaspati Srishti", Part II Charutar Vidya Mandal. Vallabh Vidyanagar, Anand, p. 413.

Bhat, S. V., Bajwa, B. S., Dornauer, H., de Souza, N. J., and Fehlhaber, H. W. (1977). *Tetrahedron Lett.* 1669.

Bhat, S. V., Dohadwalla, A. N., Bajwa, B. S., Dadkar, N. K., Dornauer, H., and de Souza, N. J. (1983). *J. Med. Chem.* **26**, 486.

Bruce, E. A. (1935). *Kew Bulletin*, 322.

Burka, J. F. (1983a). *J. Pharmacol. Exp. Ther.* **225**, 427.

Burka, J. F. (1983b). *Can. J. Physiol. Pharmacol.* **61**, 581.

Burka, J. F. (1986). *In* "Forskolin: Its Chemical, Biological and Medical Potential" (R. H. Rupp, N. J. de Souza and A. N. Dohadwalla, eds.), pp. 127–136. Hoechst India Ltd., Bombay.

Caprioli, J., and Sears, M. (1983). *Lancet* **1**, 958.

Chang, J. J., Chand, J. M., Schwalm, S., Dervinis, A., and Lewis, J. (1984). *Eur. J. Pharmacol.* **101**, 271.

Daly, J. W. (1984). *Adv. Cyclic Nucleotide Res.* **17**, 81.

de Loureiro, J. (1790). "Flora Cochinchinensis", Vol. 2, p. 272. Academy of Science, Lisbon.

de Souza, N. J. (1977). *In* "Proceedings of the Third Asian Symposium on Medicinal Plants and Spices, Colombo, Sri Lanka", pp. 83–92, UNESCO.

de Souza, N. J. (1986). *In* "Innovative Approaches in Drug Research" (A. F. Harms, ed.), pp. 191–207. Elsevier Science Publishers, Amsterdam.

de Souza, N. J., Dohadwalla, A. N., and Reden, J. (1983). *Med. Res. Rev.* **3**, 201.

Dohadwalla, A. N. (1986). *In* "Forskolin: Its Chemical, Biological and Medical Potential" (R. H. Rupp, N. J. de Souza and A. N. Dohadwalla, eds.), pp. 19–30. Hoechst India Ltd., Bombay.

Ganguli, B. N. (1986). *In* "Forskolin: Its Chemical, Biological and Medical Potential" (R. H. Rupp, N. J. de Souza and A. N. Dohadwalla, eds.), pp. 31–38. Hoechst India Ltd., Bombay.

Inamdar, P. K., Dornauer, H., and de Souza, N. J. (1980). *J. Pharm. Sci.* **69**, 1449.

Inamdar, P. K., Kanitakr, P. V., Reden, J., and de Souza, N. J. (1984). *Planta Med.* **1984**, 1.

Kaik, G. (1986). *In* "Forskolin: Its Chemical, Biological and Medical Potential" (R. H. Rupp, N. J. de Souza and A. N. Dohadwalla, eds.), pp. 137–144. Hoechst India Ltd., Bombay.

Kanatani, H., Tanimoto, J., Hidaka, K., Yamasaki, K., Kurohawa, T., and Ishibashi, S. (1985). *Planta Med.* **1985**, 182.

Khandelwal, Y., Inamdar, P. K., de Souza, N. J., Rupp, R. H., Chatterjee, S., and Ganguli, B. N. (1987a). *Tetrahedron* in press.

Khandelwal, Y., Rajagopalan, R., Dohadwalla, A. N., de Souza, N. J., and Rupp, R. H. (1987b). Unpublished data.

Kreutner, W., Chapman, R., Gulbenkian, A., and Tozzi, S. (1984). *J. Allergy Clin. Immunol.* **73**, 130A.

Laurenza, A., Khandelwal, Y., de Souza, N. J., Rupp, R. H., Metzger, H., and Seamon, K. B. (1987). *Mol. Pharmacol.* in press.

Lele, R. D., Kamdar, N. B., Popat, N., and Nair, K. G. (1986). *In* "Forskolin: Its Chemical, Biological and Medical Potential" (R. H. Rupp, N. J. de Souza and A. N. Dohadwalla, eds.), pp. 115–124. Hoechst India Ltd., Bombay.

Lichey, J., Friedrich, T., Priesnitz, M., Biamino, G., Usinger, P., and Huckauf, H. (1984). *Lancet* **2**, 167.

Linderer, T., and Biamino, G. (1986). *In* "Forskolin: Its Chemical, Biological and Medical Potential" (R. H. Rupp, N. J. de Souza and A. N. Dohadwalla, eds.), pp. 109–113. Hoechst India Ltd., Bombay.

Lindner, E., Dohadwalla, A. N., and Bhattacharya, B. K. (1978). *Arzneim.-Forsch.* **28**, 284.

Metzger, H. (1986). *In* "Forskolin: Its Chemical, Biological and Medical Potential" (R. H. Rupp, N. J. de Souza and A. N. Dohadwalla, eds.) pp. 65–80. Hoechst India Ltd., Bombay.

Metzger, H., and Lindner, E. (1981a). *IRCS Med. Sci.* **9**, 99.

Metzger, H., and Lindner, E. (1981b). *Arzneim.-Forsch.* **31**, 1248.

Metzger, H., and Lindner, E. (1982). *Hoppe Seyler's Z. Physiol. Chem.*, 66.

Mukherjee, S. K. (1940). "A Revision of the Labiatae of the Indian Empire in Records of the Botanical Survey of India", Vol. 14, No. 1, 228 pp. Government of India Press, Calcutta.

Nadkarni, S. R., Akut, P. M., Ganguli, B. N., Khandelwal, Y., de Souza, N. J., and Rupp, R. H. (1986). *Tetrahedron Lett.* **27**, 5265.

Paulus, E. F. (1980a). *Z. Krystallogr.* **152**, 43.

Paulus, E. F. (1980b). *Z. Krystallogr.* **152**, 239.

Pfeuffer, T., Gaugler, B., and Metzger, H. (1983). *FEBS Lett.* **164**, 158.

Pfeuffer, T., and Metzger, H. (1982). *FEBS Lett.* **146**, 369.

Pinto-Pereira, L. (1986). In "Forskolin: Its Chemical, Biological and Medical Potential" (R. H. Rupp, N. J. de Souza and A. N. Dohadwálla, eds.), pp. 183–191. Hoechst India Ltd., Bombay.

Ramakumar, S., Venkatesan, K., Tandon, J. S., and Dhar, M. M. (1985). *Z. Krystallogr.* **173**, 81.

Rupp, R. H., de Souza, N. J., and Dohadwalla, A. N. (eds.) (1986). "Forskolin: Its Chemical, Biological and Medical Potential". Hoechst India Ltd., Bombay.

Saksema, A. K., Green, M. J., Shue, H.-J., Wong, J. K., and McPhail, A. T. (1985). *Tetrahedron Lett.*, **26**, 551.

Seamon, K. B. (1984). *Annu. Rep. Med. Chem.* **19**, 293.

Seamon, K. B. (1985). *Drug Develop. Res.* **6**, 181.

Seamon, K. B., Padgett, W., and Daly, J. W. (1981). *Proc. Natl. Acad. Sci. U.S.A.* **78**, 3363.

Seamon, K. B., and Daly, J. W. (1981a). *J. Biol. Chem.* **256**, 9799.

Seamon, K. B., and Daly, J. W. (1981b). *J. Cyclic Nucleotide Res.* **7**, 201.

Seamon, K. B., and Daly, J. W. (1983). *Trends Pharmacol. Sci.* **4**, 120.

Seamon, K. B., and Daly, J. W. (1986). *Adv. Cyclic Nucleotide Res.* **20**, 1.

Seamon, K. B., Daly, J. W., Metzger, H., de Souza, N. J., and Reden, J. (1983). *J. Med. Chem.* **26**, 436.

Seto, C., Araie, S., Matsumoto, S., and Tekase, M. (1986). *Jpn. J. Ophthalmol.* **30**, 238.

Shah, V., Bhat, S. V., Bajwa, B. S., Dornauer, H., and de Souza, N. J. (1980). *Planta Med.* **39**, 183.

Tandon, J. S., Dhar, M. M., Ramakumar S. and Venkatesan, K. (1977). *Ind. J. Chem.*, **15B**, 880.

Wang, A. H. J., Paul, J. C., Zelnik, R., Mizuta, K., and Lavie, D. (1973). *J. Am. Chem. Soc.* **95**, 598.

Wang, A. H. J., Paul, J. C., Zelnik, R., Lavie, D., and Levy, E. C. (1974). *J. Am. Chem. Soc.* **96**, 580.

Williams, R. O. (1949). "The Useful and Ornamental Plants of Zanzibar and Pemba", pp. 47–48, 206. St. Ann's Press, Timperley, Altrincham.

Witte, P. U. (1986). *In* "Forskolin: Its Chemical, Biological and Medical Potential" (R. H. Rupp, N. J. de Souza and A. N. Dohadwalla, eds.), pp. 175–182. Hoechst India Ltd., Bombay.

2

Non-steroid, Cardioactive Plant Constituents

H. WAGNER
Institute of Pharmaceutical Biology
University of Munich
Munich, West Germany

I. INTRODUCTION

Although our present store of drugs includes very effective cardioactive agents of synthetic and biogenetic origin, the search for new active substances with a better therapeutic ratio and with different or new types of activity still continues. The recent isolation of forskolin from *Coleus forskohlii* shows that the plant kingdom offers excellent prospects in the search for further potent cardiac agents. Forskolin is a cardioactive compound with a new and hitherto unknown type of structure, displaying a new and likewise hitherto unknown specific activation of adenylate cyclase. In addition, numerous drugs used in folk medicine in many countries possess cardioactivity, but the isolation and identification of their cardioactive principles has not yet been attempted. It is now certain that systematic screening will also reveal trivial compounds, e.g., various amines that are widely

ECONOMIC AND MEDICINAL PLANT RESEARCH VOLUME 2
ISBN 0-12-730063-5

distributed in the plant kingdom, but it should not be forgotten that precisely this type of information is a prerequisite for the chemical and biological standardization of plant extracts, and this can ultimately lead to improvements in the quality of drug preparations.

Furthermore, if new types of structures are discovered, these can be used by the pharmaceutical chemist as models for the synthesis of structural analogues with improved activity.

In the present discussion, an attempt is made to differentiate between those classes of substances that can serve as a starting point for new developments, and those that simply provide scientific confirmation of hitherto empirically-used drugs, or serve for the standardization of herbal drug preparations. From the outset, it should be stated that new compounds with the potency of the cardiac glycosides are not the only goal of this search. Substances for adjuvant heart therapy, and for the geriatric heart conditions or cardiac insufficiency grades I and II (NYHA) are also of interest.

II. THE MEANING OF CARDIOACTIVE OR CARDIOTONIC

Reviews and text-books very frequently refer generally to cardiovascular action or cardiotonic action, without specifying the particular type of activity. Table I gives a list of the mechanisms of pharmacological action and the characteristic actions of all known classes of compounds of plant and synthetic origin. It can be seen that some classes of substances, like the cardiac glycosides, the sympathomimetics, or the β-receptor blockers, appear several times in the table, because they exert several different types of activity on the heart. According to pharmacological definition, the term "cardiotonic" is synonymous with "positive inotropic". In the literature, however, cardiotonic is also used to indicate an increase in frequency, an increase in the beat volume, or a general increase in cardiac performance, in addition to increased contraction. This need not always be the main type of activity, e.g., theophylline and caffeine also increase the force of contraction, but this is not a very prominent activity, and it is accepted as a welcome "side-effect".

In this review, emphasis will be placed on those compounds that show positive inotropic activity, excluding the cardiac glycosides. In addition to the known action of cardiac glycosides on membrane-bound Na^+,K^+-ATPase, at least five further mechanisms are known that may form the basis of positive inotropic activity. This must be

TABLE I
INDIVIDUAL ACTIVITIES UNDER THE COLLECTIVE TERM "CARDIAC ACTIVITY"

Activity	Definition	Active principles
Positive inotropic (= cardiotonic)	Increase of contractility	Cardiac glycosides; sympathomimetics
Negative inotropic	Decrease of contractility	β-Receptor blockers
Positive chronotropic	Increase of cardiac frequency	Sympathomimetics
Negative chronotropic	Decrease of cardiac frequency	Cardiac glycosides; β-blockers
Positive dromotropic	Increase of flow rate	Sympathomimetics
Negative dromotropic	Decrease of flow rate	Cardiac glycosides; β-blockers
Antiarrhythmic	Removal of cardiac arrhythmia	Quinidine Procainamide
Coronary dilating	Dilatation of the coronary blood vessels	Glyceryl-trinitrate Theophylline
Inhibition of thrombocytes Aggregation[a]	Prevention of a coronary infarction	Acetylsalicylic acid
Anticoagulative[a]	Prevention of a coronary infarction	Coumarin derivatives

[a]These activities have no effect on the heart, but they are quite often cited under the generic term "cardiac activity" in the literature.

taken into account in the choice of suitable *in vitro* and *in vivo* test systems when screening for cardiotonic plant constituents.

III. SCREENING AND CLASSIFICATION OF COMPOUNDS WITH POTENTIAL CARDIAC ACTIVITY

In the search for potential cardioactive compounds in plants, the following approaches can be employed in the selection of drugs for pharmacological testing.

A. Reinvestigation of old literature reports of cardiac activity.
B. Investigation of drugs used in folk medicine.
C. Selection of plants from those plant families and genera from which cardioactivity compounds have already been isolated (chemotaxonomic approach).

D. Search for further representatives of certain chemical structure types already known to possess potential cardiac activity.

Eight main classes of cardiotonic substances are known: phenylalkylamines, indole derivatives, tetrahydroisoquinolines, imidazoles and purines, certain diterpenes, sesquiterpenes, flavonoids, and other phenolic compounds.

Cardiotonic Structures

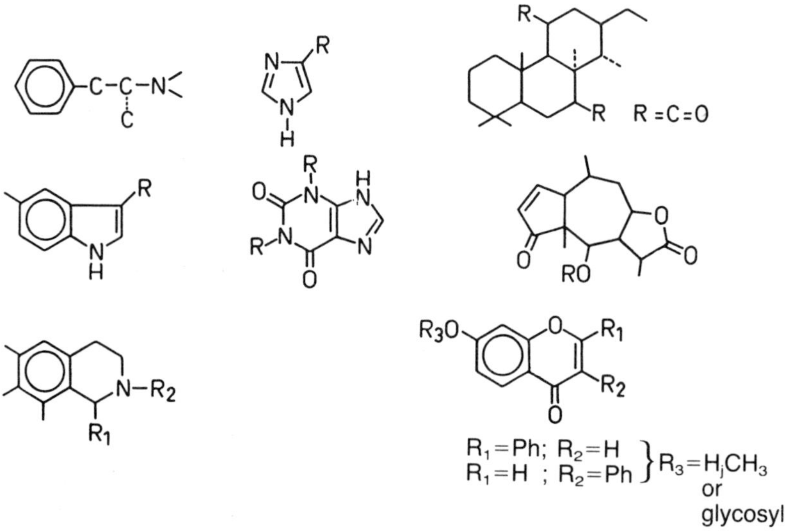

R =C=O

R_1=Ph; R_2=H ; R_1=H ; R_2=Ph } R_3=H, CH_3 or glycosyl

IV. PHENYLALKYLAMINES

The phenylpropylamines and β-phenylethylamines of this class are the longest-known non-steroid, cardioactive plant constituents. They served as model substances for the development of a great many β_1-sympathomimetic drugs. The main representative of the first type is L-ephedrine, an indirectly-acting sympathomimetic agent. Its low cardiac activity is the result of an attack on the β_1-receptors. Since ephedrine has other more prominent activities, its action on the heart is to be seen as a side-effect. In addition to ephedrine, l-ephedrine, the methylephedrines, and norephedrines, four other β-hydroxylated phenylamines have been found in plants: octopamine, synephrine (Wheaton and Stewart, 1970), macromerine from the

Phenylpropylamines, β-phenylethylamines

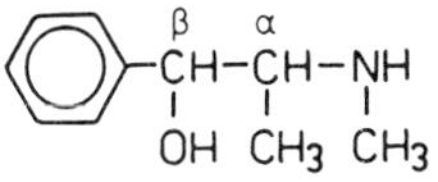

L-Ephedrine

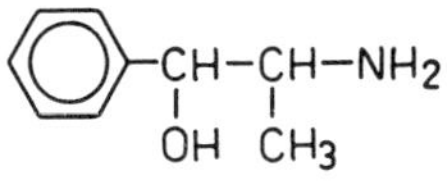

nor-ψ-Ephedrine

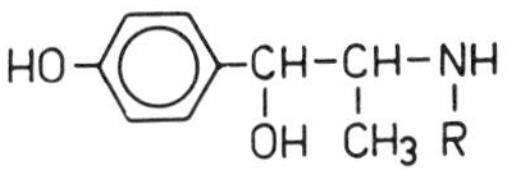

R=H: Octopamine
R=CH_3: Synephrine

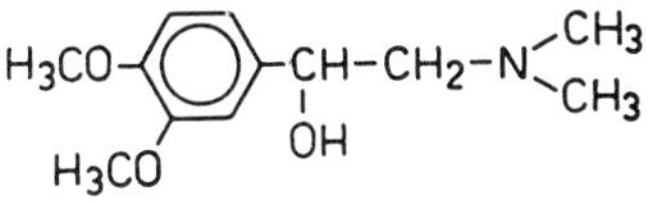

Macromerine

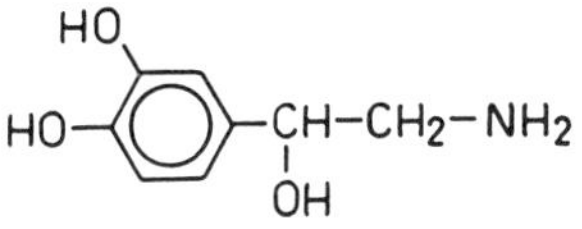

Noradrenaline

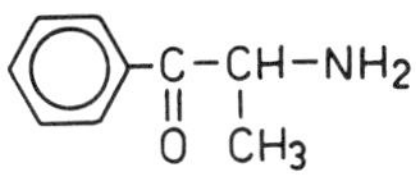

β-Aminopropiophenone
(Cathinone)

cactus *Coryphantha macromeris* (Hodgkins *et al.*, 1967), and noradrenaline (Smith, 1980). Together with the β-methoxy derivatives, they occur in members of the families, Portulacaceae, Rutaceae, Cactaceae, Amaryllidaceae, Moraceae, Musaceae, and Rosaceae (Smith, 1980). They include numerous food plants, e.g., citrus fruits, bananas, and purslane (*Portulaca oleracea*). The highest content of synephrine (up to 280 μg/g) has been reported, for example, in the fruits of *Citrus reticulata* (Smith, 1981).

β-Aminopropiophenone (cathinone), recently isolated from *Catha edulis* or "Khat" (Schorno and Steinegger, 1979), shows strong positive inotropic activity on isolated papillary muscle (Grevel, 1982). It therefore seems likely that this amine is involved in the cardiac stimulation activity of Khat leaves (Tariq *et al.*, 1984). Systematic investigations have been reported on the relationship between the pharmacological activity of this class of compound and the various types of substitution in the aromatic ring, on the nitrogen, and on the C-atoms of the side-chain (Weiner, 1980).

The second group of phenylalkylamines includes the phenylethylamines or phenethylamines, which do not possess an alcoholic hydroxy group in the β-position. They are much more widely distributed in

Positive inotropic activity of phenethylamines

	R_1	R_2	R_3	R_4	R_5	p.i. activity*
β-Phenethylamine	H	H	H	H	H	++
Tyramine	OH	H	H	H	H	++
3-Methoxytyramine	OH	OCH_3	H	H	H	+
N-Methyltyramine	OH	H	H	H	CH_3	++
Hordenine	OH	H	H	CH_3	CH_3	++
o-Methoxy-β-phenethylamine	H	H	OCH_3	H	H	+
p-Methoxy-β-phenethylamine	OCH_3	H	H	H	H	++
3,4-Dimethoxy-β-phenethylamine	OCH_3	OCH_3	H	H	H	+
Alfileramine	OH	X	H	CH_3	CH_3	−−

*++ strong positive inotropic
\+ weak positive inotropic
− negative inotropic

the plant kingdom than the ephedrine type of phenylalkylamine, and they occur with remarkable frequency in members of the Cactaceae, Rosaceae, and Rutaceae, as well as the Leguminosae (Smith, 1977). The prototype of this group is the well-studied tyramine, which, like non-substituted phenylethylamine, is an indirectly acting sympathomimetic agent, but only at concentrations below 10^{-4} *M*. At higher concentrations it shows positive inotropic activity due to a post-synaptic effect. Strong positive inotropic activity is also displayed by *N*-methyltyramine, hordenine, and *p*-methoxy-β-phenethylamine (Grevel, 1982). Using preparative ion-pair HPLC, we isolated β-phenethylamine, tyramine, and *o*-methoxy-phenethylamine from *Crataegus* flowers (Wagner and Grevel, 1982b). Tyramine, *N*-methyltyramine, and occasionally *N,N*-dimethyltyramine are found in some *Cereus* species, as well as in *Selenicereus grandiflorus*, which are still used in the form of alcoholic extracts for cardiac insufficiency and angina pectoris (Wagner and Grevel, 1982a). The digitalis-like action of *Viburnum* extracts described by Vlad *et al.* (1977) is probably due to tyramine, which has already been described by Wheaton and Stewart (1970) in *V. odoratissimum*, and by us in *V. opulus*. Tyramine and β-phenylethylamine have also been found in the drugs *Viscum album* and *Arnica montana*, which are used in heart and circulation therapy. However, cardiac action after oral administration can be expected only from those phenylalkylamines that do not act as substrates for the deactivating monoamine oxidase of the liver and intestine. This applies only to amines with a large substituent on the amino group and/or a methyl group in the β-position, e.g., as in the ephedrine compounds, β-aminopropiophenone, *N*-methylphenylethylamine, octopamine, and synephrine. On the other hand, all phenylalkylamines can be expected to show activity if they are administered parenterally. Buccal absorption of amines would probably have the same effect.

Since phenylalkylamines in a drug are relatively unstable, leading to wide variations in amine content, their role as pharmaceutically important compounds is very limited. More importance, however, must be attached to the amine content of food plants (Marquardt *et al.*, 1976). In food plants, amines can, in certain cases, give rise to undesirable side-effects, e.g., spinach contains histamine and dopamine in relatively high concentrations. Beans that are customarily eaten in New Zealand were found to cause hypertonic crises in patients with high blood pressure. This was shown to be due to the presence of L-dopa, which is converted in the body to dopamine. The juice of mandarins contains 100 μg/ml L-synephrine. Consumption of 150 ml corresponds to the minimal dose (~17 mg) prescribed for hypotonia. This represents a potential hazard, and care should be exercised, especially by patients taking monoamine oxidase inhibitors.

Cactus species have been found to contain positively inotropic histamine compounds, in addition to phenylalkylamines. The tyramine-free extract of *Echinocereus blanckii* yielded *N'*,*N'*-dimethylhistamine, as well as 3,4-dimethoxy-β-phenylethylamine (Wagner and Grevel, 1982a). Together with dolichotheline from *Dolichothele sphaerica* (Rosenberg and Paul, 1969), *N'*,*N'*-dimethylhistamine is the second imidazole derivative found so far in cacti. It possesses an EC_{50} value of 1.6×10^{-7} mol/l. A different amine, the *bis*-hordenylterpene known as alfileramine (Fig. 3), from *Zanthoxylum punctatum* (Caolo and Stermitz, 1977), surprisingly showed negative inotropic activity in our *in vitro* investigations (Grevel, 1982).

V. TETRAHYDROISOQUINOLINE (THI) DERIVATIVES

Compounds derived biosynthetically from phenylalkylamines and possessing a THI skeleton are also widely distributed in the plant

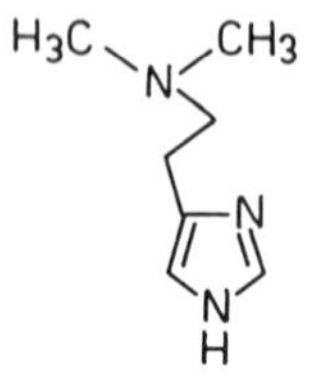

N',N'-Dimethylhistamine
(*Echinocereus blanckii*)

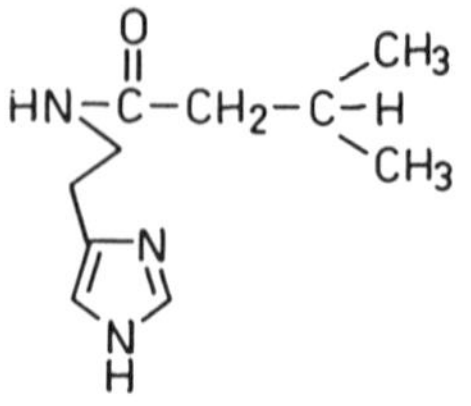

Dolichotheline
(*Dolichothele sphaerica*)

kingdom. Practically all the known representatives have been isolated from cactus species (Bruhn and Lundstrom, 1976; Mata and McLaughlin, 1980). Of the 14 THI derivatives tested, only two showed positive inotropic activity in our tests. These were peyophorine from *Lophophora williamsii* (Kapadia and Fales, 1968), and carnegine from *Carnegia gigantea* (Bruhn and Lundström, 1976). Both of these compounds are structurally related to the well-known isoproterenol. In a study by Santi-Soncin and Furlanut (1972), it was established that carnegine caused a synoidal bradycardia that was not influenced by atropine or eserine. In guinea-pig atria, carnegine showed negative chronotropic activity with a simultaneous increase in the amplitude of contraction. After prior treatment with reserpine, the inotropic effect was no longer observed, whereas chronotropic activity was increased. Thus, carnegine is an indirectly acting sympathomimetic agent. The action mechanism of peyophorine is open to conjecture. It seems likely that activity is due to stimulation of β_1-receptors, i.e., a similar action mechanism to that of the benzyltetrahydroisoquinoline, higenamine, which is discussed below. A novel discovery is that the unsubstituted aromatic compound, isoquinoline, displays positive inotropic activity. Isoquinoline was recently shown to be a

Tetrahydroisochinoline alkaloids

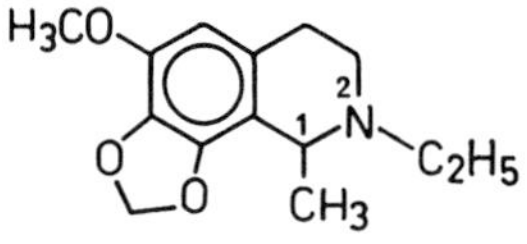

Peyophorine
(*Lophophora williamsii*)

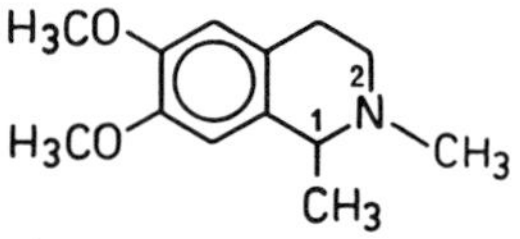

Carnegine
(*Carnegia gigantea*)

genuine constituent of the South American heart drug *Spigelia anthelmia* (Wagner *et al.*, 1986b). Isoquinoline, together with choline, benzoyl-, and 3,3-dimethylacryloyl-choline, which have been demonstrated to be in the same drug, probably contributes to the cardiovascular activity of the alcoholic extract (Wagner *et al.*, 1986a).

The presence of large substituents on C-1 has the effect of increasing the cardiac activity of tetrahydroisoquinolines. 6,7,4′-Trihydroxybenzyl-tetrahydroisoquinoline, otherwise known as demethylcoclaurine or higenamine, can be considered as the prototype of these C-1 substituted THI compounds. It was first isolated from the seeds of *Nelumbo nucifera* (Koshiyama *et al.*, 1970), then later from the roots of *Aconitum japonicum* (Bushi) (Kosuge and Yokota, 1976), and from the leaves of *Annona squamosa* (Wagner *et al.*, 1980). Higenamine possesses positive inotropic activity at a concentration of about 10^{-7} g/ml, i.e., similar to that of isoproterenol. The catecholamine-like activity arises from an attack on β-receptors, and

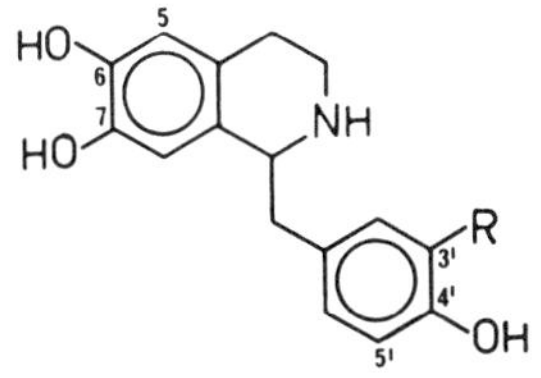

R=H : Higenamine
R=OH : Tetrahydropapaveroline

the β_2-sympathomimetic activity is more pronounced than the β_1-activity (Wagner *et al.*, 1980). Detailed pharmacological investigations of higenamine were carried out by Park *et al.* (1984) on the left atrium of the rabbit. Phytochemical and pharmacological studies by Konno *et al.* (1979) with an extract of *Aconitum carmichaeli*, also known by the collective name of "Bushi", have shown, however, that higenamine is apparently not solely responsible for the cardiotonic action of *Aconitum* root. They found that the main alkaloid, mesaconitine, and another isolated compound, coryneine (dopamine methochloride), are involved in the cardiotonic activity. The cardiotonic action of naturally occurring 1-benzyl-tetrahydroisoquinoline derivatives (BTHI) was not unexpected, because it had been known for a long time that tetrahydropapaveroline (THP) possesses similar activity (Holtz *et al.*, 1964; Nott and Head, 1978; Brossi *et *al.*, 1980). Tetrahydropapaveroline can arise from dopamine in the mammalian organism in the presence of monoamine oxidase, and it is found in the urine of Parkinson patients treated with L-dopa. New systematic investigations in the 1-benzyl-THI series were performed by Miller *et al.* (1975), Piascik *et al.* (1978), and Ikezawa *et al.* (1978a,b). They found that the *o*-dihydroxy grouping in positions 6 and 7, which is present in higenamine and tetrahydropapaveroline, is not necessarily a prerequisite for β-sympathomimetic activity. The monohydroxy and the dihydroxy compounds showed similar activities on isolated tracheal strips, whereas the latter compound showed markedly higher activity on the atrium of the guinea-pig. Tetrahydropapaveroline, with two catechol moieties, is also more active than higenamine. Compounds methoxylated at positions 3′, 4′, and 5′ of the benzyl moiety were likewise found to possess good, but weaker activity. The 3′, 4′, 5′-trimethoxybenzyl derivative, with two OH-groups in position 7, merely showed graded effects on bronchial and heart muscle, depending on the state of untreated tonus.

Interestingly, the 6-*O*-methylated BTHI, e.g., *O*-methylarmepavine, *N-nor-O*-methylarmepavine, *N*-norarmepavine, or *N*-methylcoclaurine, showed negative inotropic activity in our investigations.

VI. NITROGEN-CONTAINING COMPOUNDS WITH OTHER CARBON SKELETONS

Cardiotonic activity is also found among representatives of the aporphine, berberine, and benzophenanthridine series, which are biosynthetically closely related to the BTHI compounds. Alkaloids from the bark of *Cymbopetalum brasiliense* (Annonaceae) act synergistically, and are at least partly responsible for the positive inotropic activity of the aqueous drug extract (Cave *et al.*, 1984). The isolated

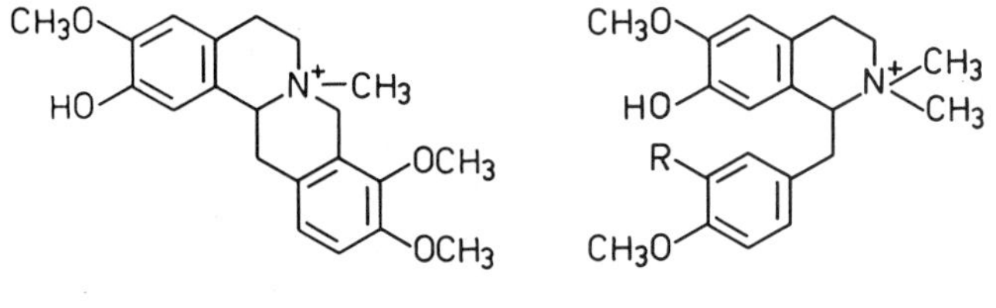

N-Methylisocorypalmine

R=OH Tembetarine
R=H Colletine

alkaloids, *N*-methylisocorypalmine, tembetarine, and colletine, however, were active on the isolated perfused rat heart only in relatively high doses (~5 mg).

In the berberine series, methylcanadine from *Zanthoxylum coriaceum* (Swinehart and Stermitz, 1980), and sanguinarine, isolated earlier from *Sanguinaria canadensis*, possess positive inotropic activity. Sanguinarine exerts its positive inotropic action by inhibition of Na^+,K^+-ATPase (Akera *et al.*, 1981).

Among those compounds not biosynthetically related to phenylalanine or dopamine, there are several that have been known for

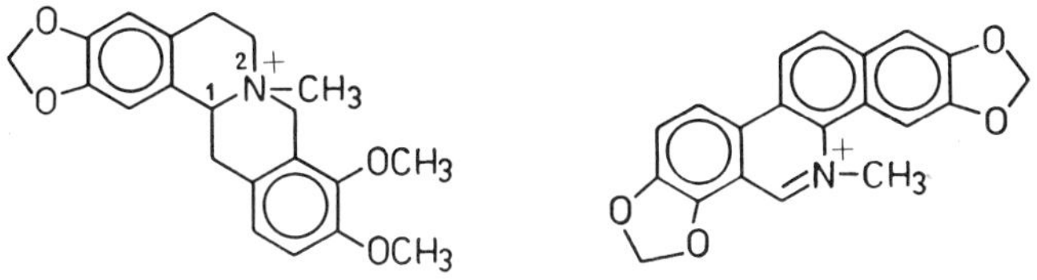

N-Methylcanadine
(*Zanthoxylum coriaceum*)

Sanguinarine
(*Sanguinaria canadensis*)

a long time. The lupine alkaloid sparteine possesses specific antiarrhythmic activity, similar to that of quinidine. This positive inotropic effect, measured on isolated guinea-pig heart muscle, cannot, however, be demonstrated in the whole animal. In the narcotized dog, intravenous injection of 1 mg/kg even has a negative inotropic effect (Raschak, 1974). The diuretic action described for extracts of *Sarothamnus scoparius* and *Spartium junceum* is presumably due to the presence of the flavone C-glycoside scoparin (Steinegger and Hänsel, 1968).

As already mentioned, the cardiotonic action of theophylline (Fig. 9) resulting from an inhibition of phosphodiesterase is considered to be a side-effect. This compound is frequently combined with cardiac glycosides for therapeutic purposes. The related compound, cyclic AMP (Fig. 9), likewise possesses inotropic properties. Cyclic AMP is widely distributed in the plant kingdom, as recently shown by radioimmunoanalysis and bioluminescence assay (Brown and Newton, 1981). In view of the low concentrations found so far (2–1000 pmol/g fresh weight), a pharmacological role for cAMP-containing plant extracts is excluded, even with parenteral administration. The same would seem to apply to adenosine and 2′-

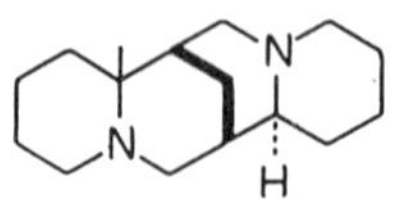

Sparteine

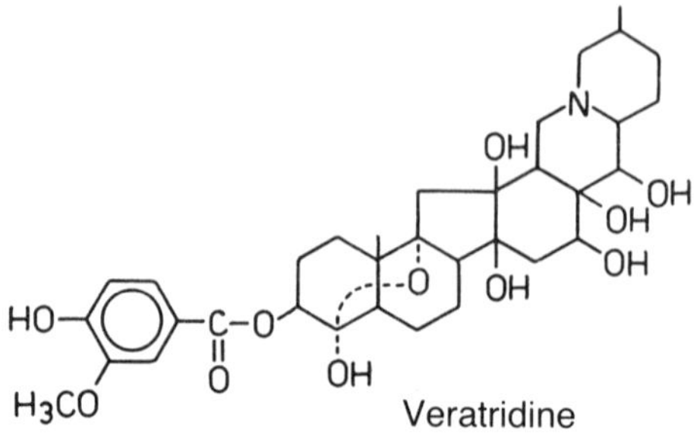

Veratridine

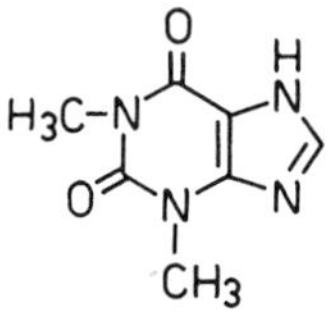

Theophylline

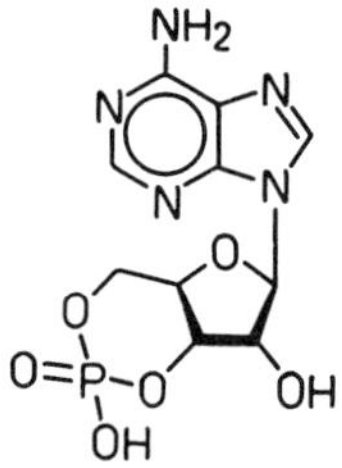

cyclo-AMP

deoxyadenosine. Adenosine has been found in some plants, e.g., in the onion (Weisenberger *et al.*, 1972), garlic (Michahelles, 1974), and *Crataegus* (Fiedler *et al.*, 1958). Hypoxanthine-9-L-arabinofuranoside has been isolated from root extracts of *Boerhaavia diffusa* (Ojewole and Adesina, 1985), but this extract is not recognized as a cardiac agent in folk medicine.

VII. CARDIOTONIC PLANT CONSTITUENTS WITH TERPENOID STRUCTURES

Compounds in the first group, e.g., the *Erythrophleum* alkaloids, possess a diterpene structure, and they also contain heterocyclic nitrogen. *Erythrophleum* is a genus of the Leguminosae found on the west coast and inland in equatorial Africa. *Erythrophleum guineense* is a tree resembling a stone pine, and an extract of its bark, which is cardiotoxic, is used by the natives as an arrow poison (Neuwinger, 1974). All the *Erythrophleum* alkaloids known so far are diterpene carboxylic acids, mostly esterified with dimethylaminoethanol (Hauth, 1971, 1974). As shown below for cassaine, a semicyclic double bond at position 13 is a component of the α,β-unsaturated

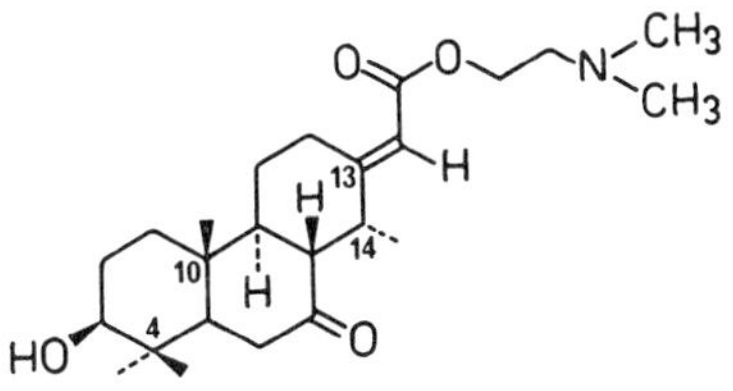

Cassaine

ester grouping. In analogy with steroids, a CH_3 group is present at position 10. One CH_3 group is also present at position 14, and at least one is found at position 4. The alkaloids differ from one another by the nature of the substitution at positions 3, 4, 6, and 7. The positive inotropic activity of the main alkaloid is the result of binding to Na^+,K^+-ATPase. Compared with the cardiac glycosides, however, their affinity for the enzyme and therefore their inotropic activity are lower (Akera *et al.*, 1981). After it had been recognized that the α,β-unsaturated diethylaminobenzyl ester structure was essential for activity, and that the C=O group on C–7 did not greatly influence activity, derivatization studies were undertaken, especially with regard to the C–3 OH. Some derivatives, e.g., the acetylamide or glucuronylamide, were found to be superior to cassaine with respect to both an increase and the duration of the minute volume (Hauth, 1971). It was therefore surprising to find that a nitrogen-free, relatively simple diterpene, lacking the diethylamino group, displayed positive inotropic activity. However, the mechanism of action of forskolin from the roots of *Coleus forskohlii*, a labdane-type diterpene, is totally different from that of the cardiac glycosides, veratrine, or theophylline. Forskolin is an adenylate cyclase inhibitor, showing a dose-dependent activation of the cAMP-dependent protein kinase of

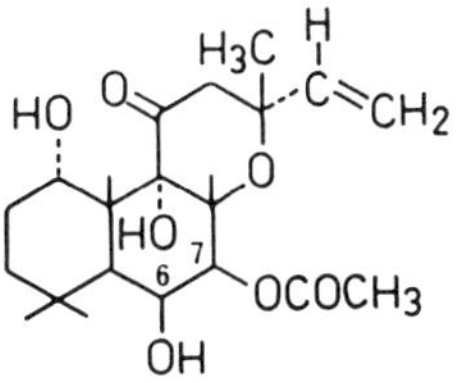

Forskolin
(*Coleus forskohlii*)

heart muscle. Active concentrations lie in the range of 10 n*M* and higher. For details of the pharmacology, the reader should consult Chapter 1. As a result of the discovery of forskolin, other plant drugs were systematically investigated for the presence of adenylate cyclase activators of this type (Kanatani *et al.*, 1985). Fifteen drugs showed a marked stimulation of adenylate cyclase. In comparison with 28% for *Coleus forskohlii*, activities ranged from 30 to 163% (see Table II).

TABLE II
DRUGS SHOWING MARKED ADENYLATE CYCLASE ACTIVITIES

Armeniacae semen (30%)	Scutellariae radix (83%)
Atractylodis Lanceae rhiz. (31%)	Perillae herba (82%)
Atractylodis rhizoma (39%)	Menthae herba (59%)
Platicodi radix (39%)	Swertiae herba (53%)
Sinomoni cauris rhizoma (42%)	Skizonepetae herba (44%)
Bupleuri radix (70%)	Forsythiae fructus (43%)
Pharbitidis semen (163%)	Zedoariae rhizoma (40%)
Gambir (87%)	

The identities of the structures responsible for this activity and their relative specificities are unknown.

The grayanotoxins or andromedotoxins from *Andromeda japonica* are also diterpenes, displaying positive inotropic activity in isolated heart preparations, as well as *in vivo*. In the case of aconitine and the toad poison batrachotoxin, activity is due to a prolongation of Na^+-channel opening time (Akera *et al.*, 1981). The action mechanism of two positively inotropic quassinoids, ailanthone and dihydroailanthone, is unknown (Grevel, 1982). The activity, however, appears

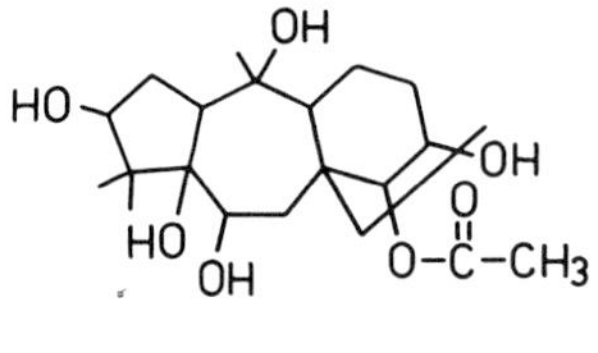

Grayanotroxin (Andromedotoxin)
(*Andromeda japonica*)

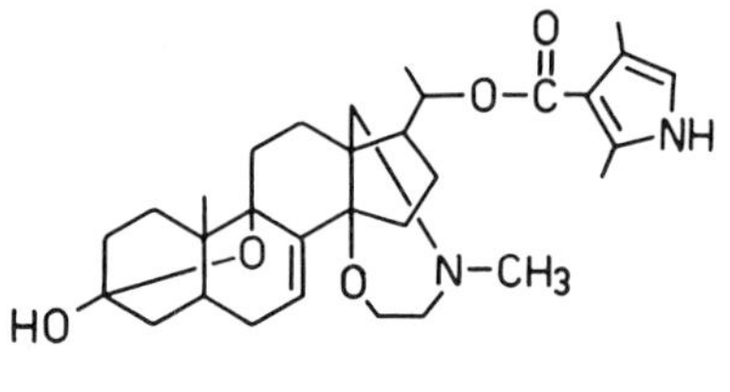

Batrachotoxin

to confirm earlier studies by Weger (1929) on the action of the bitter principles, quassin, artemisin, and some extracts of bitter principle drugs on isolated heart muscle. Although *Arnica* flowers have been used for a long time as a heart stimulant and mild analeptic, it was only recently that Willuhn and Kresken (1981) demonstrated that the sesquiterpene lactones helenalin, helenalin acetate, and dihydrohelenalin acetate are responsible for the cardiotonic and the cardiotoxic properties of this drug. Helenalin acetate, at a concentration of 10^{-6} to 10^{-4} M, showed positive inotropic activity on the isolated spontaneously-beating right atrium and on isolated intact papillary muscle of the guinea-pig (Willuhn, 1981).

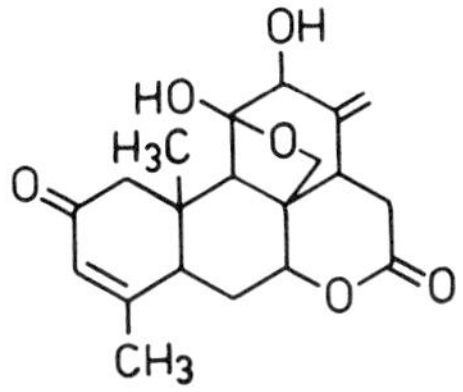

Ailanthone
(*Ailanthus glandulosa*)

Higher concentrations (orally 123 mg/kg; i.p. 31 mg/kg) caused toxic symptoms in mice. The results are in agreement with earlier investigations on helenalin by Lamson (1913). The cardiotonic activity may be due to inhibition of cAMP-phosphodiesterase, which has already been demonstrated in Ehrlich ascites tumour cells. In this connection, mention must also be made of the cardiotoxic cucurbitacin glycoside, gradiotoxin from *Gratiola officinalis* (Müller and Wichtl, 1979). According to investigations by Wichtl, "gradiotoxin" is very probably identical to cucurbitacin E-glucoside. On Langendorff

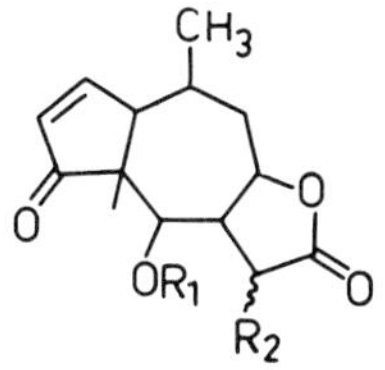

R_1:H; R_2:$=CH_2$ Helenalin
R_1:CH_3CO; R_2:$=CH_2$ Helenalin acetate
R_1:CH_3CH; R_2: CH_3 Dihydrohelenalin acetate

hearts, in doses of 10^{-6} to 10^{-3} g/ml of perfusion solution, the cardiac activity of the cucurbitacins is characterized by negative inotropy and chronotropy, and by increased coronary flow. In common with the terpenes mentioned above, the cucurbitacins possess α,β-unsaturated ketone groups. All these compounds are also cytotoxic and taste bitter. The joint occurrence of these two properties and cardiotonic activity is striking. According to Kupchan *et al.* (1971), the α,β-unsaturated lactone ring of the cardiac glycosides is responsible for all three of these properties. The same hypothesis may therefore assist the search for cardiotonic terpenoids.

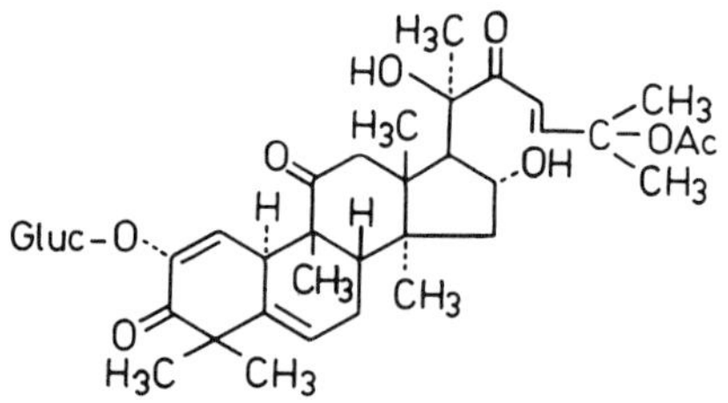

Cucurbitacin E-glucoside
(=Elaterinide=Gradiotoxin) (*Gratiola offic.*)

VIII. FLAVONOIDS AND OTHER PHENOLIC COMPOUNDS

Although there have been numerous investigations on the cardiotonic action of flavonoids (Böhm, 1967), their therapeutic usefulness is limited to a few synthetic isoflavonoids and the flavonoid-containing *Crataegus* extract. Measured on Langendorff hearts, ipriflavone (= 7-isopropoxyisoflavone) decreases the oxygen consumption of cardiac muscle, in both hypoxia and ischaemia, and its activity is predominantly that of an oxygen-sparing agent (Feuer *et al.*, 1981). Some carboxymethylene ethers of 7-hydroxyisoflavone (e.g. Recordyl[R]) possess positive inotropic and coronary dilatatory activity. Studies on the relationship between this activity and the substitution type show that the positive inotropic activity is greatest for the 7-*O*-substituted flavones if the C-2 position is occupied by only H or CH_3. Larger residues decrease the activity to the point where the introduction of a second phenyl residue results in loss of activity (Dinya and Hetenyi, 1975). The action mechanism of this

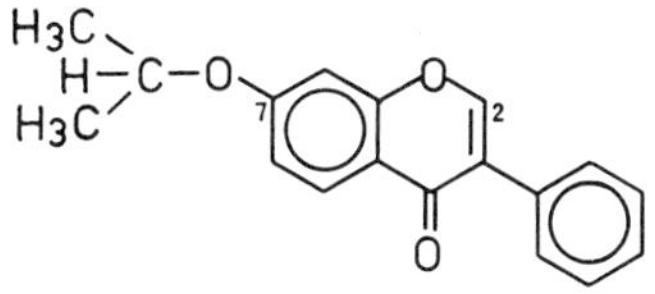

7-Isopropoxyisoflavone
(Ipriflavone)

positive inotropic activity is unknown. A β-receptor blocking action appears to be excluded, but activity is conceivably due to inhibition of phosphodiesterase. Comprehensive investigations and reviews have been published on the cardioactivity of *Crataegus* extracts. In a double-blind study conducted in Japan by Iwamoto *et al.* (1981), it was shown that *Crataegus* extracts are active at the onset of cardiac insufficiency grades I and II on the NYHA scale. As shown by Weinges *et al.* (1971), Kukovetz (1976), and Gabard and Trunzler (1983), *Crataegus* extracts and an oligoprocyanidin concentrate from *Crataegus* show positive inotropic activity on the Langendorff heart. Furthermore, *Crataegus* extract also increases coronary flow in the myocardial circulation; it shows positive chronotropic, positive dromotropic, and negative bathmotropic activity; it increases tolerance of the myocardium to oxygen deficiency, increases the heart–time volume, decreases peripheral vessel resistance, and increases overall

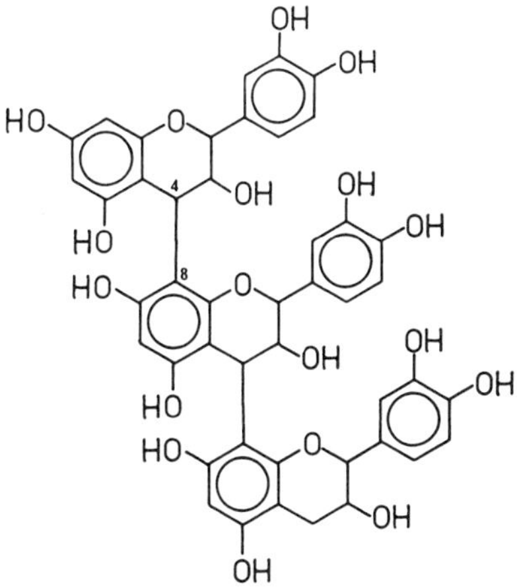

Procyanidin trimer

cardiac performance (Gabard and Trunzler, 1983); the precise mechanism of action is unknown. The main active principles are thought to be the procyanidin oligomers (Weinges *et al.*, 1971; Rewerski and Lewak, 1967; Rewerski *et al.*, 1971), although the amines present in the "phenylalkylamine" fraction must also be considered as possible active agents.

There is considerable evidence that the described flavonoids exert their cardiotonic action at least partly by inhibition of cellular phosphodiesterase and elevation of the cellular cAMP concentration, as well as by affecting the permeability of cell organelles to Ca^{2+} ions. Systematic investigations with pure flavonoids were carried out by Beretz *et al.* (1978, 1979, 1986), Petkov *et al.* (1981), and Nikaido

TABLE III
INHIBITION OF 3′,5′-AMP-PHOSPHODIESTERASE BY FLAVONOIDS

Flavonoids	IC_{50} (μM)
1. *Flavonols*	
Kämpferol	5.2
Quercetin	5.5
Rutin	7.2
Quercitrin	100.0
2. *Flavones*	
Luteolin	4.3
Apigenin	9.2
Vitexin	7.1
Luteolin-7-glucoside	10.0
Biflavonoids	
Rhusflavone [=naringenin-(I-6, II-8)-apigenin]	1.0
Biapigenins	
I-3,II-6 Robustaflavone	4.0
I-3′,II-6 Robustaflavone	4.0
I-8,II-8 Cupressuflavone	4.9
I-6,II-8 Agathisflavone	1.0
I-3′,II-8 Amentoflavone	0.12
3. *Flavones*	
Naringenin	50.0
Naringin	10.0

et al. (1982). According to these studies, inhibitory activity decreased in the order flavonol, flavone, anthocyanidin, flavanone, flavanonol, catechin (Table III). Polymethoxyflavonoids were stronger inhibitors than the corresponding polyhydroxy compounds (Nikaido *et al.*, 1982). Some biflavonoids also showed good activity, i.e., rhusflavone and especially the biapigenins amentoflavone and agathisflavone (Table III) (IC_{50} = 0.12–1.0 μM) (Chakravarthy *et al.*, 1981; Beretz *et al.*, 1979).

Flavone mixtures isolated from *Ruta graveolens*, *Prunus spinosa*, *Rosa canina*, *Crataegus oxyacantha*, and *Vaccinium myrtillus* were as active as the most powerfully active single compounds (Petkov *et al.*, 1981). This inhibitory activity towards phosphodiesterase is not limited to flavonoid structures. Nikaido *et al.* (1981) found that a series of lignans were also potential phosphodiesterase inhibitors. According to a review by MacRae and Towers (1984), active compounds are found in both the 1-phenyl-tetrahydronaphthalene and the dioxybicyclooctane series. Since the diglucoside of pinoresinol from the traditional Chinese drug *Eucommia ulmoides* possesses antihypertensive activity (Sih *et al.*, 1976), it is conceivable that the cardiotonic action of other lignan-containing drugs, e.g., mistletoe (Wagner *et al.*, 1986a), owe their activity to their constituent lignans.

A new discovery is that some sharp-tasting phenols, e.g., yakuchinone-A[1(4′-hydroxy-3′-methoxyphenyl)-7-phenyl-3-heptanone]from *Alpinia oxyphylla* (Shoji *et al.*, 1984) and the 6,8- and 10-gingerols from *Zingiber officinale* (Shoji *et al.*, 1982), possess strong positive inotropic activity on isolated guinea-pig atrium. The action mechanism is assumed to be similar to that of the cardiac glycosides, i.e., inhibition of the Na^+,K^+-pump.

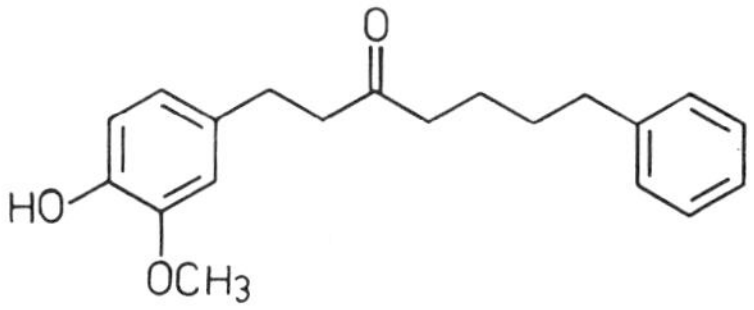

O
HO
OCH3

Yakuchinone-A

IX. DISCUSSION AND SUMMARY

The search for new cardiotonically active non-steroid compounds in plants is justified as follows. First, the many existing cardiac glycosides are known to have a relatively small therapeutic ratio. Second, one must consider the poor prospects of finding new cardiac glycosides that retain a high therapeutic efficacy and a large therapeutic index. The case of forskolin demonstrates quite well that such a search can be quite successful if appropriate screening techniques are employed. Currently, there are eight classes of non-steroid-containing compounds with positive inotropic activities: phenylalkylamines, indole derivatives, tetrahydroisoquinolines, imidazoles, purines, diterpenes, sesquiterpenes, and flavonoids.

The properties of the most important compounds, their origins, and their mechanisms of action have been described. Individual substances and corresponding drugs have been discussed with a view towards therapeutic applications. With the exception of purines, the tetrahydroisoquinolines and diterpene compounds seem to be very attractive candidates for structural design studies aimed at improving their therapeutic efficacy. Other structures usually are less cardioactive. However, the cardioactive properties of these compounds may be sufficiently high to warrant their use in the treatment of grade I and II cardiac insufficiency (NYHA), as is demonstrated, e.g., by flavonoids, procyanidins, and *Crataegus* extracts. In addition, these compounds also offer, for the first time, the possibility of chemical standardization and optimization of drug preparations that have been used empirically in traditional medicine for a long time.

With respect to screening strategies, it is important to differentiate between positive inotropic activities and other activities, such as antiarrhythmic, dilatatory, or sympathomimetic. This information is required in order to select suitable *in vitro* and *in vivo* test systems.

There are currently only very few reports available that describe the absorption, bioavailability, and metabolism of non-steroid cardiotonic compounds. Future pharmacological studies, therefore, will have to focus on these areas of investigation.

ACKNOWLEDGEMENTS

The author thanks Dr. Grevel and Dr. Seitz for their assistance in preparing this manuscript.

REFERENCES

Akera, T., Fox, A. L., and Greeff, K. (1981). *In* "Handbuch der Experimentellen Pharmakologie", p. 459. Springer Verlag, Berlin.

Beretz, A., Anton, R., and Cazenavo, J. P. (1986). *In* "Plant Flavonoids in Biology and Medicine", Proceedings of a symposium held in Buffalo, New York, July 22–26, 1985 (V. Cody, E. Middleton and J. B. Harborne, eds.), p. 281. Alan R. Liss, Inc., New York.'

Beretz, A., Anton, R., and Stoclet, J. C. (1978). *Experientia* **34**, 1054.

Beretz, A., Joly, M., Stoclet, J. C., and Anton, R. (1979). *Planta Med.* **36**, 193.

Böhm, K. (1967). "Die Flavonoide". Čantor, KG, Aulendorf/Wttbg.

Brossi, A., Rice, K. C., Mak, Ch.-P., Reden, J., Jacobson, A. E., Nimitkitpaisan, Y., Skolnick, Ph., and Daly, J. (1980). *J. Med. Chem.* **23**, 648.

Brown, E. G., and Newton, R. P. (1981). *Phytochemistry* **20**, 2453.

Bruhn, J. G., and Lundström, J. (1976). *Lloydia* **39**, 197.

Caolo, M. A., and Stermitz, F. R. (1979). *Tetrahedron* **35**, 1487.

Cave, A., Debourges, G., Lewin, G., Moretti, C., and Dupont, Ch. (1984). *Planta Med.* **50**, 517.

Chakravarthy, B. K., Rao, Y. U., Gambhir, S. S., and Gode, K. D. (1981). *Planta Med.* **43**, 64.

Dinya, Z., and Hetenyi, E. (1975). *In* "Topics in Flavonoid Chemistry and Biochemistry (L. Farkasm, M. Gabor, and F. Kallay, eds.), p. 240. *Academiai Kiado*, Budapest.

Feuer, L., Barth, P., Stráuss, I., and Kékes, E. (1981). *Arzneim.-Forsch.* **31**, 953.

Fiedler, U., Hildebrand, G., and Neu, R. (1958). *Arzneim.-Forsch.* **3**, 436.

Gabard, B., and Trunzler, G. (1983). *In* "Wandlungen in der Therapie der Herzinsuffizienz" (N. Rietbrock, B. Schmiedes, and J. Schuster, eds.). Verlag Friedr. Vieweg & Sohn, Braunschweig.

Grevel, J. (1982). "Chemisches und pharmakologisches Screening zum Auffinden nichtsteroider herzwirksamer Pflanzenwirkstoffe", Thesis, Munich.

Hauth, H. (1971). *Planta Med. Suppl.* **4**, 40.

Hauth, H. (1974). *Planta Med.* **25**, 201.

Hodgkins, J. E., Brown, S. D., and Massingill, J. L. (1967). *Tetrahedron Lett.* 1321.

Holtz, P., Stock, K., and Westermann, E. (1964). *Naunyn-Schmiedebergs Arch. Exp. Pathol. Pharmakol.* **248**, 387.

Ikezawa, K., Narita, H., Ikeo, T., Umino, N., Iwakuma, T., and Sato, M. (1978a). *Folia Pharmacol. Jpn.* **74**, 663.

Ikezawa, K., Takenaga, H., Ivie, K., Sato, M., Nakajima, H., and Kiyomoti, A. (1978b). *Folia Pharmacol. Jpn.* **74**, 819.

Iwamoto, M., Ishizaki, T., and Sato, T. (1981). *Planta Med.* **42**, 1.

Kanatani, H., Tanimoto, J., Hidaka, K., Kohda, H., Yamasaki, K., Kurokawa, T., and Ishibashi, S. (1985). *Planta Med.* **2**, 182.

Kapadia, G. J., and Fales, H. M. (1968). *J. Pharm. Sci.* **57**, 2017.

Konno, Ch., Shirasaka, M., and Hikino, H. (1979). *Planta Med.* **35**, 150.

Koshiyama, H., Ohkuma, H., Kaaguchi, H., Hsu, H. Y., and Chen, Y. P. (1970). *Chem. Pharm. Bull.* **18**, 2564.

Kosuge, T., and Yokota, M. (1976). *Chem. Pharm. Bull.* **24**, 176.

Kukovetz, W. (1976). *Pharm. Ztg.* **121**, 1429.

Kupchan, S. M., Eakin, M. A., and Thomas, A. M. (1971). *J. Med. Chem.* **14**, 1147.

Lamson, P. D. (1913). *J. Pharmacol. Exp. Ther.* **4**, 471.

MacRae, W. D., and Towers, G. H. N. (1984). *Phytochemistry* **23**(6), 1207.

Marquardt, P., Classen, H. G., and Schumacher, K. A. (1976). *Arzneim.-Forsch.* **26**, 2001.
Mata, R., and McLaughlin, J. L. (1980). *Phytochemistry* **19**, 673.
Michahelles, E. (1974). "Uber neue Wirkstoffe aus Knoblauch (Allium sativum L.) und Küchenzwiebel (Allium cepa L.)", Thesis, Munich.
Miller, D. D., Merritt, W. V. P., Kador, P. F., and Feller, D. R. (1975). *J. Med. Chem.* **18**(1), 99.
Müller, A., and Wichtl, A. (1979). *Pharm. Ztg.* **124**, 1761.
Neuwinger, H. D. (1974). *Naturwiss. Rundsch.* **27**, 385.
Nikaido, T., Ohmoto, T., Noguchi, H., Kinoshita, T., Saitoh, H., and Sankawa, U. (1981). *Planta Med.* **43**, 18.
Nikaido, T., Ohmoto, T., Sankawa, U., Hamanaka, T., and Totsuka, K. (1982). *Planta Med.* **46**, 162.
Nott, M. W., and Head, G. A. (1978). *Clin. Exp. Pharmacol. Physiol.* **5**, 313.
Ojewole, J. A. O., and Adesina, S. K. (1984). *Fitoterapia* **56**, 31.
Park, C. W., Chang, K. C., and Lim, J. K. (1984). *Arch. Int. Pharmacodyn. Ther.* **267**, 279.
Petkov, E., Nikolov, N., and Uzunow, P. (1981). *Planta Med.* **43**, 183.
Piascik, M. T., Osei-Gyimah, P., Müller, D. D., and Feller, D. R. (1978). *Eur J. Pharmacol.* **48**, 393.
Raschack, M. (1974). *Arzneim.-Forsch.* **24**, 753.
Rewerski, W., and Lewak, S. (1967). *Arzneim.-Forsch.* **17**, 490.
Rewerski, W., Piechocki, T., Rylski, M., and Lewak, S. (1971). *Arzneim.-Forsch.* **21**, 886.
Rosenberg, H., and Paul, A. G. (1969). *Tetrahedron Lett.* 1039.
Santi-Soncin, E., and Furlanut, M. (1972). *Fitoterapia* **43**, 21.
Schorno, X., and Steinegger, E. (1979). *Experientia* **35**, 572.
Shoji, N., Iwasa, A., Takemoto, T., Ishida, Y., and Ohizumi, Y. (1982). *J. Pharm. Sci.* **71**, 1174.
Shoji, N., Umeyama, A., Takemoto, T., and Ohizumi, Y. (1984). *Planta Med.* **56**, 186.
Sih, C. J., Ravikumar, P. R., Huang, F. C., Buckner, C., and Whitlock, H. I. (1976). *J. Am. Chem. Soc.* **98**, 5412.
Smith, T. A. (1977). *Phytochemistry* **16**, 9.
Smith, T. A. (1981). *Food Chem.* **6**, 169.
Steinegger, E., and Hänsel, R. (1968). "Lehrbuch der Pharmakognosie auf phytochemischer Grundlage", Vol. 2. Springer-Verlag, Berlin, Heidelberg, New York.
Swinehart, J. A., and Stermitz, F. R. (1980). *Phytochemistry* **19**, 1219.
Tariq, M., Ageel, A. M., Parmar, N. S., and Al-Meshal, I. A. (1984). *Fitoterapia* **55**, 195.
Vlad, L., Munta, A., and Crisan, J. G. (1977). *Planta Med.* **31**, 228.
Wagner, H., Feil, B., Seligmann, O., Petricic, J., and Kalogjera, Z. (1986a). *Planta Med.* **48**, 102.
Wagner, H., and Grevel, J. (1982a). *Planta Med.* **45**, 95.
Wagner, H. and Grevel, J. (1982b). *Planta med.* **45**, 98.
Wagner, H., Reiter, M., and Ferstl, W. (1980). *Planta Med.* **40**, 77.
Wagner, H., Seegert, K., Gupta, M. P., Esposito Avella, M., and Solis, P. (1986b). *Planta Med.* **48**, 378.
Weger, P. (1929). *Naunyn-Schmiedebergs Arch. Exp. Pathol. Pharmakol.* **144**, 261.
Weiner, N. (1980). *In* "The Pharmacological Basis of Therapeutics" (L. S. Goodman and A. Gilman, eds.), p. 138. MacMillan, New York.
Weinges, K., Kloss, P., Trunzler, G., and Schuler, R. (1971). *Planta Med. Suppl.* **4**, 61.
Weisenberger, H., Grube, H., Koenig, E., and Pelzer, H. (1972). *FEBS Lett.* **26**, 105.
Wheaton, T. A., and Stewart, I. (1970). *Lloydia* **33**, 244.
Willuhn, G. (1981). *Pharmazie in Unserer Zeit* **10**, 1.
Willuhn, G., and Kresken, J. (1981). *Planta Med.* **42**, 107.

3

Natural Products for Liver Diseases

HIROSHI HIKINO
YOSHINOBU KISO
Pharmaceutical Institute
Tohoku University
Sendai, Japan

I. INTRODUCTION

Many diseases have now been counteracted by the improvement of sanitary conditions and the development of drugs, especially antibiotics and vaccines. However, a great number of people who suffer chronic diseases still are without appropriate therapies. One such disease concerns liver disorders.

Steroids have been mainly employed for the therapy of hepatitis. Although steroids exhibit curative effects to a certain extent in many cases, their adverse effects cannot be ignored, particularly in chronic diseases, because they are administered for long terms.

Recently, emphasis has been placed on the development of vaccines and antiviral drugs. In fact, HB vaccine has been developed, and interferon and its inducers have been confirmed to have effectiveness in some cases of viral hepatitis. However, HB vaccine may be used only for the prevention and not for the cure of HB viral hepatitis, and also it cannot be used for other types of viruses. Although viruses are the main cause of liver disease in certain countries, liver lesions induced by hepatotoxic chemicals (especially ethyl alcohol,

ECONOMIC AND MEDICINAL PLANT RESEARCH VOLUME 2
ISBN 0-12-730063-5

peroxides (particularly peroxidized edible oil), toxins in food (especially aflatoxins), pharmaceuticals (mainly antibiotics, chemotherapeutics, CNS-active drugs), environmental pollutants, etc., are also serious problems. Although liver lesions produced by hepatotoxins may be recovered by cessation of their ingestion at early stages, they cannot be healed only by removal of the toxins after critical periods. Therefore, really good remedies for liver diseases are needed.

On the other hand, there are a number of natural drugs that have been claimed to have curative effects on liver disorders in traditional medicine. Recent progress in the study of traditional drugs has resulted in the isolation of a number of active principles, including antihepatotoxic constituents. Their pharmacological and biochemical actions have been examined and the clinical effects of some of them have been evaluated for liver protective activity on a scientific basis, particularly using double-blind tests. There are some drugs that originally were utilized for other purposes but later were empirically revealed to possess liver protective effects.

This chapter is an attempt to summarize recent advances in natural drugs for liver diseases. Although a tremendous amount of work has been conducted in this field, only the results of recent investigations on the most important natural drugs used for the treatment of liver diseases will be discussed, because of space limitations. Previous results are only mentioned briefly, because they have been the object of excellent reviews (Koch, 1980; Wagner, 1981). A comprehensive survey of liver-protective drugs of plant origin has recently been published (Handa *et al.*, 1986).

II. SILYMARIN

The application of extracts of the milk thistle, *Silybum marianum* Gaertn. (Compositae), as a liver drug is not the result of a systematic screening but rather is due to its clinical history of use in traditional medicine. It was shown from historical records that decoctions of *Silybum* fruits were utilized by the Greeks as early as B.C.

Following the isolation of silymarin from the seeds of *S. marianum* (Wagner *et al.*, 1968), a scientific basis of this plant for the treatment of liver diseases, which had previously been employed in the form of inexactly-defined galenical preparations, was established. Early pharmacology investigations proved that silymarin reduced the harmful actions on the liver of carbon tetrachloride (CCl_4), thioacetamide, α-amanitin (Hahn *et al.*, 1968), and phalloidin (Vogel, 1968; Vogel and Temme, 1969). From knowledge that was acquired at that stage, it was believed that silymarin was a single substance of the flavonol group (Wagner *et al.*, 1968), but later it was shown to be a mixture of three components, silybin, silydianin, and silychristin, belonging to the class of 2-phenylchromanones of the taxifolin type, containing a molecule of coniferyl alcohol in oxidative binding (Wagner *et al.*, 1974).

Earlier studies demonstrated a protective effect of oral silymarin administration against CCl_4-induced liver damage in rats. Indeed silymarin was also found to reduce the prolongation of hexobarbital sleeping time produced by CCl_4 and to prevent the inhibition of hepatic metabolism of *p*-oxyphenylpyruvic acid induced by CCl_4 (Hahn *et al.*, 1968). Carbon tetrachloride intoxication raised the serum levels of various enzymes such as glutamic–oxalacetic transaminase (GOT), glutamic–pyruvic transaminase (GPT) and sorbitol dehydrogenase (SDH), but under treatment with silymarin these increases were significantly diminished (Rauen and Schriewer, 1971, 1973; Schriewer *et al.*, 1973a,c; Vogel *et al.*, 1975).

The intraperitoneal injection of D-galactosamine (GalN) to rats causes an acute hepatitis that is functionally and histologically similar to viral hepatitis in humans (Keppler *et al.*, 1968). Silymarin was investigated and was shown to have a protective action against liver lesions induced by GalN at low toxic doses (Rauen and Schriewer, 1971; Schriewer and Rauen, 1973) but not at higher toxic doses (Meyer-Burg, 1972).

Poisoning with DL-ethionine leads to accumulation of triglycerides in the liver of rats. It was found that silymarin inhibited increases of triglycerides by DL-ethionine in the liver (Vazquez de Prada, 1974).

Thioacetamide is a hepatotoxin, which shows toxic activity when repeatedly administered to rats. Giving it in the feed every day for several months causes a development of conditions that are somewhat like the liver cirrhosis clinically observed in humans. When silymarin was given along with long-term administration of thioacetamide in the feed, animals lost less weight and their survival times increased significantly (Hahn *et al.*, 1968). A single injection of thioacetamide to rats leads to increases in the serum levels of the enzymes GPT, GOT, SDH, and glutamate dehydrogenase, which were also prevented by silymarin (Schriewer *et al.*, 1973b).

It is now established that fatty liver due to ethyl alcohol poisoning in rats is brought about by increased biosynthesis of fatty acids,

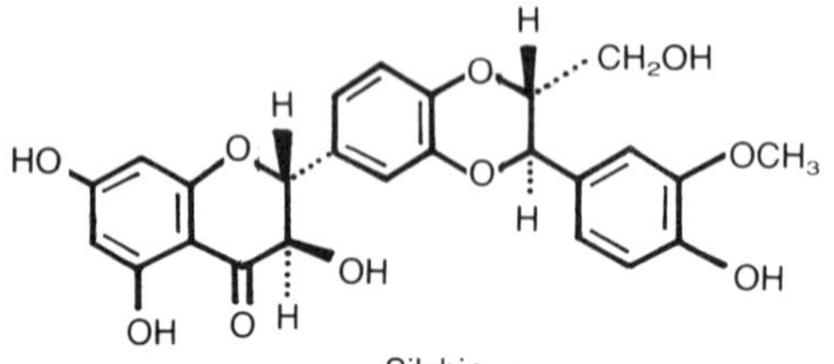

Silybin

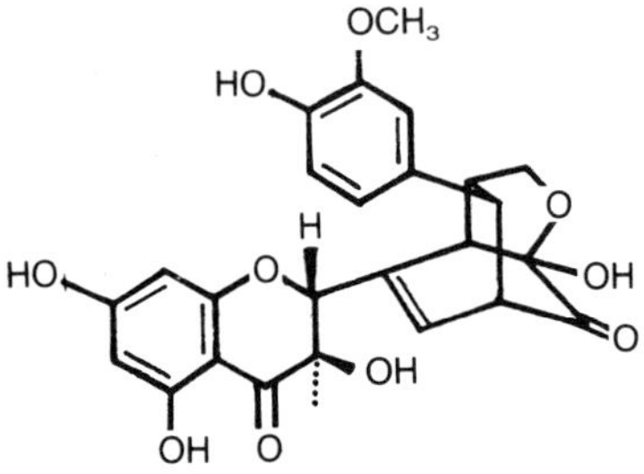

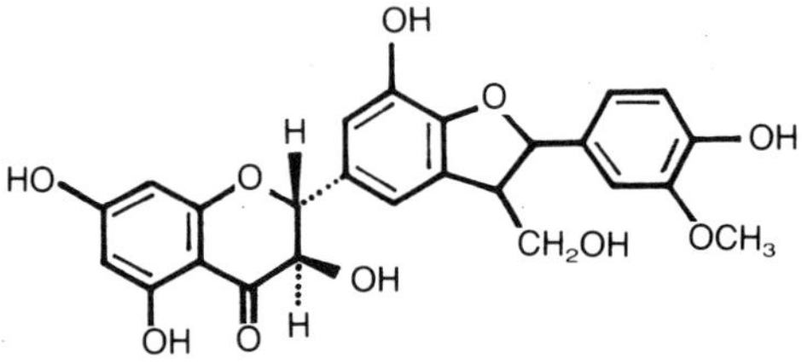

Silydianin

Silychristin

increased esterification of fatty acids to triglycerides, and decreased oxidation of fatty acids in mitochondria. The biochemical and electronoptical changes in mitochondria caused by alcohol are almost completely prevented by silymarin (Platt and Schnorr, 1971). Ethyl alcohol treatment produces marked increases in malondialdehyde formation and in spontaneous chemiluminescence in the rat liver. When silymarin was given prior to acute ethyl alcohol treatment, it was possible to suppress the increases of both of these effects completely. Possible mechanisms that may be responsible for the inhibition of ethyl-alcohol-induced lipid peroxidation by silymarin pretreatment include scavenging of free radicals and increases in the hepatic contents of both reduced and oxidized glutathione (Valenzuela *et al.*, 1985).

In all experimental models used for liver damage tests, the one most definitely antagonized by silymarin is the damage induced by the toxins from the poisonous mushroom *Amanita phalloides*, i.e., phalloidin and α-amanitin. Thus, silymarin showed protective and curative effects on survival time and death rate of mice after administration of α-amanitin (Hahn *et al.*, 1968). Silymarin also antagonized the lethal toxicity of phalloidin when it was administered either before or after the toxin in mice. The antihepatotoxic potency of silymarin against phalloidin was stronger than that against α-amanitin (Vogel, 1968; Vogel and Temme, 1969). Antiphalloidin activity of silymarin was also demonstrated in histochemical and histoenzymological studies (Desplaces *et al.*, 1975). However, it was observed that its effect depended on the interval between the poisoning with phalloidin or α-amanitin and the administration of silymarin. When silymarin was administered later than 20 min after poisoning, it was no longer possible to detect any antihepatotoxic effect (Braatz, 1976; Desplaces, 1976). From the time dependency of its effect against the two *Amanita* toxins, it was suggested that silymarin prevents penetration of the toxins by competing with the toxins for the same receptor on cell membranes.

Apart from various *in vivo* investigations on silymarin, performed on model systems of liver lesions, a comparative screening of six *Silybum* flavonolignans was carried out by assaying *in vitro* for antihepatotoxic activity using CCl_4- and GalN-induced cytotoxicity in primary cultured rat hepatocytes as model systems. Remarkable inhibitory actions were observed with silybin, silandrin, silymonin, and 3-deoxysilychristin in CCl_4-treated cultures, while silydianin and silymonin effectively prevented GalN-induced cell lesion (Hikino *et al.*, 1984b).

In vitro studies with nuclei from rat liver cells demonstrated that silybin increased the incorporation of [^{3}H]uridine triphosphate into liver RNA and enhanced ribosomal RNA synthesis as a result of the stimulation of DNA-dependent RNA-polymerase A (Machicao and Sonnenbichler, 1977; Sonnenbichler and Pohl, 1980). Influence of silibinin (formerly expressed as silybindihemisuccinate, which gave better solubility) on DNA synthesis in normal rat livers and in livers from partially hepatectomized rats was investigated. As representative of fast-growing cells, rat and human hepatoma cells and HeLa- and Burkitt lymphoma cell cultures were examined concurrently. In hepatectomized liver, a remarkable increase in DNA synthesis caused by silibinin was observed and no effect was found in the case of normal livers and in the malignant cell lines (Sonnenbichler *et al.*, 1986). These results give additional proof for the liver cell regenerating capacity of the flavonolignan derivative, silibinin, supporting clinical reports (Fintelmann and Albert, 1980).

Up to the early 1970s silymarin was shown to have remarkable therapeutic effects, not only in toxic and metabolic liver damage (Schriefers and Dietz, 1969; Schopen and Lange, 1970; Schmidt, 1971) but also in hepatitis; both acute (Wilhelm and Haase, 1973) and chronic cases (Holzgartner, 1970; Hammerl *et al.*, 1971; Milosavljevic and Eckenberg, 1972; Alvisi *et al.*, 1974).

The therapeutic effect of silymarin on acute viral hepatitis was further evaluated by a double-blind study. Thus, the values of serum bilirubin, GOT, and GPT in 28 patients treated with silymarin were compared with those in 29 patients treated with a placebo. The parameters in the silymarin-treated group were more improved than those in the control group after the 5th day of treatment. The number of patients having attained normal values after 3 weeks' treatment was higher in the silymarin-treated group than in the control group. A statistical comparison revealed a significant difference in the bilirubin and GOT values between the silymarin and control groups, and a definite trend in the regression of the GPT values in favour of silymarin. However, the course of the immune reaction in hepatitis B surface antigen (HBsAg) was not influenced by silymarin (Magliulo *et al.*, 1978).

The curative effect of silymarin on toxic liver damage was also tested in 33 patients treated with silymarin in comparison with placebo control in a double-blind trial; the cause of the toxic damage was not taken into account, but in most cases it was alcohol. The parameters GOT, GPT, and γ-glutamyltranspeptidase (γ-GTP) were improved significantly by silymarin, sometimes returning to normal in a much shorter time than in the control group. The result of this trial showed that silymarin was superior to the placebo (Fintelmann and Albert, 1980).

Examination of the therapeutic effect of silymarin was carried out using biochemical and morphological alterations in the liver as the indexes in a double-blind controlled study. Thus, 106 patients having diseased liver, mostly induced by alcohol abuse, were selected on the basis of elevated serum transaminase levels and randomly allocated into the silymarin-treated and control groups. Decreases of serum GPT and GOT in the treated group were statistically significant over those of the controls. Significant improvement of patients in the treated group were also observed in the bromosulphophthalein (BSP) retention test and histological examination. The concentrations of total and conjugated serum bilirubins were reduced more in the treated than in the controls, although the differences were not statistically significant (Salmi and Sarna, 1982).

The effect of silymarin in comparison with a placebo was assessed in a double-blind test with 21 patients suffering from chronic hepatic diseases, when 20 out of 21 patients were observed during 12 months; 1 patient had to be excluded at the end of the first month because of adverse effects of nausea and vomiting. Silymarin proved to be better than the placebo as regards the prevention of decrease of seralbumin from 3 months of therapy to the end of the study. The histopathological findings of focal necrosis and fibrosis were much improved in the patients treated with silymarin as compared with those treated with placebo after 6 months of medication. Silymarin proved to be beneficial in the parameters of parenchymal disorders, intralobular mesenchymal reaction, and fibrosis, at the end of the test (Berenguer and Carrasco, 1977).

Contrary to the above findings, two double-blind studies in which the effects of silymarin on chronic hepatitis were investigated with 24 and 12 patients, respectively, gave negative results on the parameters examined and did not exhibit any significant differences between silymarin and placebo treatment. According to the histological changes, however, silymarin treatment was sometimes significantly superior to placebo treatment (Kiesewetter *et al.*, 1977).

Long-term treatment with silymarin on chronic hepatopathies caused by psychopharmaceuticals resulted in significant improvements in the liver function parameters such as GOT, GPT, and BSP during treatment, as compared with the control group (Saba *et al.*, 1976).

These basic and clinical data suggest that silymarin is a therapeutically useful drug that stabilizes the cell membrane, stimulates protein synthesis, and accelerates the process of regeneration. These are considered to play important roles in the therapeutic efficacy of silymarin.

III. GLYCYRRHIZIN

The crude drug "kanzō" from the root of licorice, *Glycyrrhiza glabra* L. and its varieties, and from Chinese licorice, *G. uralensis* Fischer (Leguminosae), is a quite well-known drug in the East and West.

In Oriental medicine, its main use is to sweeten decoctions, to mitigate the action of drastic drugs, and to relieve pain caused by muscle contraction. In Europe, it has been employed as a corrective, an expectorant, an antitussive, and an antiulcer agent.

Glycyrrhizin is one of the main components of licorice root. It consists of one molecule of glycyrrhetinic acid and two molecules of glucuronic acid. Because glycyrrhizin contains glucuronic acid, it was first assumed that glycyrrhizin might have an antidotal action. Based on this viewpoint, a preparation of glycyrrhizin has been used for allergies and detoxification in the field of dermatology in Japan since 1948. During the course of such clinical use, glycyrrhizin preparations were empirically found to be effective for chronic hepatitis and have been widely used for chronic hepatitis and liver cirrhosis in Japan. Therefore, unlike other substances, the basic data on the liver-protective activity of glycyrrhizin accumulated after the clinical findings. Because the general pharmacological actions of glycyrrhizin have been outlined in a previous review (Hikino, 1985), this chapter will only summarize data on the antihepatotoxic actions of glycyrrhizin and the therapeutic uses of glycyrrhizin preparations.

COOH
H
O
RO

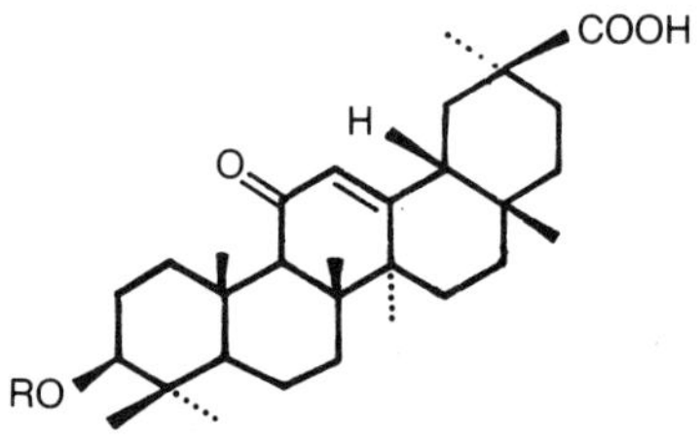

Glycyrrhizin: R = D-Glc A-β (1→2)-D-Glc A
Glycyrrhetinic acid: R = H

Examination of the effect of glycyrrhizin on GalN-induced liver damage demonstrated that glycyrrhizin inhibited liver cell injury but did not reverse reduced protein synthesis. Uridine not only improved morphological lesions but also reduced protein synthesis (Okita *et al.*, 1975). Additionally, on GalN-induced liver damage, glycyrrhizin has been confirmed to have inhibitory actions against CCl_4-, benzene hexachloride, mitomycin C-, and polychlorinated biphenyl-induced liver lesions by means of histological observations (Watari, 1972, 1973, 1975, 1977; Watari and Torizawa, 1972). Recently, using primary cultured rat hepatocytes, it was found that glycyrrhizin showed protective action against hepatocytotoxicity induced by CCl_4 (Nakamura and Ichihara, 1981; Kiso *et al.*, 1983a) and by GalN (Kiso *et al.*, 1983b). Investigation of the structure–activity relationships in glycyrrhizin and its analogues in CCl_4- and GalN-induced cell damage indicated that, among the analogues, glycyrrhetinic acid and glycyrrhizin exhibited their most prominent effects on cytotoxicity induced by each hepatotoxin (Kiso *et al.*, 1984). Much evidence has been accumulated that CCl_4 is first metabolized by drug-metabolizing enzymes (represented by cytochrome P-450 in the liver endoplasmic reticulum) to the highly reactive CCl_3 radical, which induces lipid peroxidation, covalent binding to macromolecules, and inhibition of the calcium pump of microsomes, leading to liver damage (Recknagel, 1983). The mechanism of inhibitory activity of glycyrrhizin and glycyrrhetinic acid against CCl_4-induced liver lesions was investigated, demonstrating that the antihepatotoxic action of glycyrrhizin and glycyrrhetinic acid is not dependent on its steroidal action because the artificial 18α-glycyrrhetinic acid, which has an oxygen function on the E-ring, thus adopting a similar conformation to that of prednisolone, revealed significantly stronger anti-inflammatory activity than that of the natural 18β-glycyrrhetinic acid, which exhibited much less antihepatotoxic action. Further, from the results obtained by determining the effects of these substances on the initial steps in the sequence of events leading to CCl_4-induced liver lesions, it was suggested that the antioxidative actions of glycyrrhizin and glycyrrhetinic acid participate in the antihepatotoxic activity (Kiso *et al.*, 1984).

Immunological mechanisms are now recognized to play an important role in the induction and progression of human liver disease. A model of experimental chronic active hepatitis in rabbits has been produced by immunization with human liver specific lipoprotein (LSP), which is a plasma-membrane-associated antigen complex (Meyer zum Buschenfelde and Miescher, 1972; Meyer zum Buschenfelde *et al.*, 1972; Meyer zum Buschenfelde and Hopf, 1974), and anti-LSP antibody was detected in the sera of immunized animals (Hopf *et al.*, 1974) and also in the sera of patients with various liver diseases (Hopf *et al.*, 1976; Kakumu *et al.*, 1979; Manns *et al.*, 1980). Further studies have shown that the level of anti-LSP antibody correlated with the magnitude of liver injury (Jensen *et al.*, 1978), indicating that anti-LSP antibody may be involved in the pathogenesis of liver disease. Although these observations suggest one possibility that patients with chronic active hepatitis are in a state hypersensitive to LSP, problems remain to be solved concerning how such a hypersensitive state results in liver injury.

Antibody-dependent cell-mediated cytotoxicity (ADCC) has been considered to take part in liver cell damage in hepatitis. Indeed, using isolated rabbit hepatocytes as target cells, it has been demonstrated that lymphocyte cytotoxicity for hepatocytes is observed in non-T cell populations from patients with chronic active hepatitis, and is blocked by the addition of LSP (Cochrane *et al.*, 1976). Further, it has also been suggested that liver cell damage may be induced by activated macrophages infiltrating into the liver tissue of patients, because various lymphokines, including macrophage activating factor, are produced by peripheral blood lymphocytes (PBL) from patients with liver diseases, especially chronic active hepatitis, when stimulated with LSP (Mizoguchi *et al.*, 1981a,b). In order to investigate the effect of glycyrrhizin on immunologically induced liver damage, an *in vitro* study using isolated rat liver cells was employed. Thus, an ADCC reaction was induced by co-culture with isolated liver cells coated with anti-LSP antibody and peripheral blood mononuclear cells as effector cells. When the culture supernatant separated from the ADCC reaction mixture and was added to freshly prepared liver cells, protein synthesis in liver cells was reduced. Pretreatment of liver cells with glycyrrhizin prevented the reduction of protein synthesis significantly, but not in a dose-dependent manner (Mizoguchi *et al.*, 1985a). Similarly, activated macrophage-mediated cytotoxicity was produced by addition of cultured supernatant of peritoneal exudate macrophages, activated by lipopolysaccharide, to freshly prepared isolated liver cells. Reduced protein synthesis was observed in the supernatant-treated culture. When the liver cells were pretreated with glycyrrhizin, reduction of protein synthesis was significantly prevented but with limited doses. The most effective concentration of glycyrrhizin in ADCC and activated macrophage-mediated cytotoxicity was 20 μg/ml, and excess amounts of glycyrrhizin were not necessarily beneficial. From these results, it was concluded that glycyrrhizin may protect liver cells from immunologically induced liver injuries (Mizoguchi *et al.*, 1985a).

The enhancement of antibody production by glycyrrhizin has been investigated by *in vitro* methods. When mononuclear cells from human peripheral blood were stimulated with pokeweed mitogen (PWM) in the presence of glycyrrhizin, polyclonal antibody production was significantly enhanced as compared with the case in which mononuclear cells were stimulated with PWM alone. Because the culture supernatant from glycyrrhizin-treated monocytes also caused an enhancement of PWM-induced antibody response, the enhancement of antibody production was attributed, at least partially, to the activation of monocytes. By gel filtration, the active component contained in the culture supernatant of glycyrrhizin-treated monocytes was assumed to be interleukin 1 because the fractionated substance was similar in molecular size to interleukin 1 and caused an increase in DNA synthesis of phytohaemagglutinin(PHA)-stimulated thymocytes. These findings suggest that glycyrrhizin may have an enhancing effect on antibody formation through the production of interleukin 1 (Mizoguchi *et al.*, 1984a).

The possibility of interferon induction by glycyrrhizin was examined using mice, demonstrating that two phases of the induction of interferon in serum takes place; the induced interferon was regarded as interferon γ (Abe, N., *et al.*, 1982). This induction may be followed by activation of macrophages (Schultz *et al.*, 1977) and augmentation of natural killer (NK) activity through the action of the induced interferon (Djeu *et al.*, 1979).

Investigation of the effect of glycyrrhizin on virus growth showed that glycyrrhizin inhibited the growth of several DNA and RNA viruses but had no effect on protein synthesis and replication of uninfected cells. In addition, glycyrrhizin inactivated herpes simplex virus particles irreversibly (Pompei *et al.*, 1979).

The glycyrrhizin preparation (SNMC) presently employed for clinical purposes in Japan consists of glycyrrhizin (40 mg), cysteine (20 mg), and glycine (400 mg) in 20 ml of physiological saline.

In 1958, the first report appeared announcing that SNMC administration improved altered liver function of patients with chronic hepatitis, especially as shown by the BSP retention test and thymol turbidity test (Yamamoto *et al.*, 1958). Since then, the clinical effect of SNMC to improve serum transaminase abnormalities in chronic hepatitis and liver cirrhosis has been observed in successive trials. In particular, the therapeutic effect of SNMC against chronic hepatitis has been confirmed by conducting a double-blind test with 133 patients by intravenous injection of SNMC at a dose of 40 ml (glycyrrhizin, 80 mg)/day for 30 consecutive days. In liver function tests, elevated serum transaminase and γ-GTP levels were markedly reduced by treatment with SNMC after medication for 1 week and 4 weeks, respectively, with high statistical significance. Improvement in serum transaminase levels seemed to play a major role in the evaluation of a general usefulness rating. No marked side-effects, such as pseudoaldosteronism, were observed under the conditions employed throughout the medication period. Thus, SNMC was concluded to be effective for the treatment of chronic liver diseases (Suzuki *et al.*, 1983). Recently, a higher dose of SNMC was administered to chronic hepatitis patients, leading to the conclusion that administration of the i.v. dose of 100 ml (glycyrrhizin, 200 mg)/day gave better results, without marked side-effects, than that of 40 ml in a short period (Hino *et al.*, 1981; Kumada *et al.*, 1983). Long-term administration of SNMC (40 ml/day) to patients with an HBe antigen positive test caused a change to HBe antigen negative at the higher rate of 11 out of 18 patients (61%), and a change to HBe antibody positive in the remaining 7 patients (39%) (Fujisawa *et al.*, 1980, 1984).

Although several reports have recently appeared dealing with the efficacy of glycyrrhizin on the pretreatment of post-transfusion hepatitis, a double-blind, randomized, controlled trial was conducted to compare the relative efficacies of SNMC and inactive placebo in 336 patients. In comparison with the placebo group, a significant reduction of the incidence of non-B hepatitis after transfusion was observed in the SNMC-treated group (40 ml/day) from the day of initial transfusion to the 14th day after final transfusion. Because a remarkable reduction of the incidence of post-transfusion hepatitis was observed from 2 weeks to 6 weeks after transfusion, it was suggested that the incidence of short-incubation post-transfusion hepatitis would be effectively suppressed by SNMC administration (Suzuki *et al.*, 1980). Further, SNMC was tested for its prevention of post-transfusion hepatitis in surgical clinics. Thus, when a 40 ml/day intravenous administration was continued for about 2 weeks, starting on the day of transfusion, the incidence of hepatitis was apparently reduced from 17.6 to 12.8% (Kasai *et al.*, 1981). The preventive effect was more significant in recent studies where the dose of SNMC was increased from 40 to 60 ml/day, and from 60 to 100 ml/day. When 91 patients were given 100 ml SNMC/day by intravenous drip, from the first transfusion day until 1 week after the final transfusion (mean period of administration, 2 weeks), the incidence of hepatitis 18 weeks after transfusion was only 3.6%, showing a significant difference from that (17.6%) in patients not receiving SNMC. From these results, it was concluded that the use of this drug is effective for the prophylaxis of post-transfusion hepatitis (Sekiguchi *et al.*, 1982).

As a result of the wide clinical use of large doses of glycyrrhizin, an adverse effect called pseudoaldosteronism, causing symptoms such as hypertension, oedema, and hypokalaemia, became significant (Molhuysen *et al.*, 1950). Although this adverse effect could be reduced by cessation of glycyrrhizin administration or concurrent dosing of an antialdosterone agent such as spironolactone, continuous use of glycyrrhizin was discontinued in some cases. As a result of one of the efforts to separate the therapeutic actions from the side-effects, an analogue, olean-12-en-3β,30-diol, was prepared which experimentally exhibited lower toxicity than glycyrrhizin, exerted no potentiation of aldosterone, but still possessed certain favourable effects (Takahashi *et al.*, 1980). Efforts in this direction may lead to successful preparation of clinically promising candidates.

From the results of a number of trials examining antihepatotoxic and therapeutic effects of glycyrrhizin in liver lesions, it is considered that glycyrrhizin is a useful drug for liver diseases, although an ambiguity still remains because the therapeutic efficacy of glycyrrhizin may be dependent on the lowering effect of serum transaminase levels.

IV. (+)-CATECHIN

(+)-Catechin is one of the flavonoids isolated from *Uncaria gambir* Roxburgh (Rubiaceae) and certain other plants, and known in commerce as (+)-cyanidanol-3 or cianidanol. It is a component of a number of condensed tannins, and in itself possesses astringent activity and, therefore, has been utilized for medicinal, dyeing, and tanning purposes.

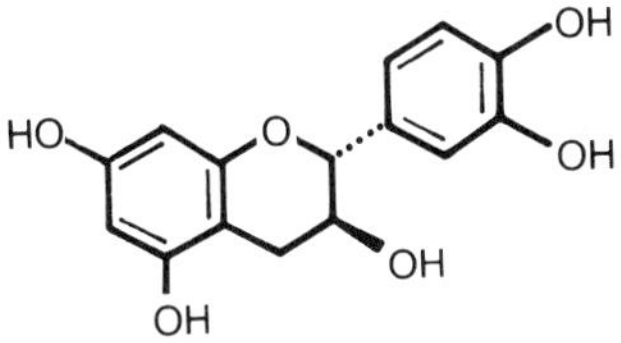

HO O OH OH OH OH

(+)-Catechin

Up to now, (+)-catechin has been shown to have hepatoprotective effects against a number of hepatotoxic agents such as ethionine (Gajdos *et al.*, 1970a; Bertelli, 1975), ethyl alcohol (Gajdos *et al.*, 1970b, 1972a,b, 1975; Ryle *et al.*, 1981; Videla *et al.*, 1981), GalN (Reutter *et al.*, 1975; Perrissoud and Weibel, 1980), CCl_4 (Bertelli, 1975; Perrissoud and Weibel, 1980; Sato *et al.*, 1983), α-naphthylisothiocyanate (Bertelli, 1975), hyperbaric oxygen (Bertelli, 1975), paracetamol (McLean and Nuttall, 1978), phalloidin (Perrissoud *et al.*, 1981), and acetaminophen (Yamada *et al.*, 1981) in experimental animals. The mode of action is uncertain but is thought to involve free-radical scavenging (Slater and Eakins, 1975; Slater and Scott, 1981), antioxidant (Kappus *et al.*, 1979; Koster-Albrecht *et al.*, 1979), and membrane-stabilizing properties (Hennings, 1979).

It has been shown that (+)-catechin is able to favour the elimination of hepatitis B surface antigen (HBsAg) from the blood of patients with acute HB hepatitis (Blum *et al.*, 1977), suggesting that (+)-catechin might stimulate the cell-mediated immune response because cellular immunity is now known to play an important part in the elimination of virus. The possible action of (+)-catechin on cellular immunity has been investigated by leukocyte-migration inhibition tests using leukocytes that were from normal donors and HBsAg positive patients, and sensitized with tuberculin-purified protein derivative (PPD) and HBsAg, respectively. Although (+)-catechin did not modify leukocyte migration in the absence of an antigen, at a concentration of 25 μg/ml it amplified the inhibition of migration by 7% in normal donor leukocytes sensitized with PPD and by 10.5% in leukocytes from HBsAg positive patients when sensitized with HBsAg. It is thus considered that (+)-catechin amplified the cell-mediated immune response and thus eliminated HBsAg (Vallotton and Frei, 1981). In addition, (+)-catechin has been shown to normalize the reduced T-lymphocyte number in patients with chronic liver disease (Sipos *et al.*, 1980) and to accelerate PPD-induced lymphocyte transformation in tuberculin positive patients with chronic active liver disease (Yamamoto, S. *et al.*, 1981). These observations led to the hypothesis that (+)-catechin could generate populations of cells capable of regulating human immune response.

Further studies have been undertaken to determine the effect of (+)-catechin on the activity of mouse cytotoxic T-lymphocytes (CTL) and NK cells. It was found that the CTL activity of the spleen cells from mice sensitized with EL-4 cells were augmented by oral administration of (+)-catechin in a dose-dependent manner (Ikeda *et al.*, 1984a). Augmentation of NK cell activity by (+)-catechin has also been demonstrated using YAC-1 cells (Ikeda *et al.*, 1984b).

Suppressor T-cell function of patients with acute viral hepatitis is transiently enhanced, whereas that of patients with chronic active hepatitis is decreased when assessed by PWM-stimulated immunoglobulin (Ig) production and by concanavalin A (Con A)-stimulated blast transformation of PBL (Kakumu *et al.*, 1980; Kashio *et al.*, 1981). Studies have been reported to determine if (+)-catechin could activate or induce suppressor cells to depress Ig synthesis and blast transformation of PBL obtained from normal humans and patients with chronic active liver disease. Thus, (+)-catechin added directly to cultures of PWM-stimulated PBL or co-cultures of B- and T-cells from normal individuals caused severe suppression of Ig production, which was mediated by radiosensitive T-cells. Since it is known that such radiosensitive T-cells belong to suppressor T-cells (Siegal and Siegal, 1977), this T-cell subset may be selectively activated by (+)-

catechin. When normal T-cells pre-incubated with (+)-catechin (25 ug/ml) for 48 hr were cultured with freshly prepared autologous or allogenic normal PBL in the presence of PWM, Ig production was markedly suppressed. Similarly, there was a consistent suppression of blast transformation of Con-A-stimulated autologous or allogenic responding PBL by normal T-cells pretreated with (+)-catechin. On the other hand, (+)-catechin-induced suppressor activity of T-cells from patients with chronic active liver disease was significantly lower than that of normal individuals. Based on these results, it was alleged that altered suppressor activity in liver disease may be corrected by the administration of (+)-catechin (Kakurnu *et al.*, 1983).

The intensity of picryl-chloride-induced delayed-type hypersensitivity (PCI-DTH) was significantly lowered in mice with experimental liver injury induced by the injection of CCl_4 or α-naphthylisothiocyanate, as compared with normal mice. Adherent spleen cells prepared from mice with liver injury suppressed the PCI-DTH in normal syngenic mice by adoptive transfer at the time of immunization with picryl chloride. (+)-Catechin, administered orally to recipient mice immediately after the transfer of adherent spleen cells, caused total prevention of the PCI-DTH from suppression by the transferred adherent spleen cells. In an *in vitro* study, the adherent spleen cells from both CCl_4-treated and α-naphthylisothiocyanate-treated mice produced higher amounts of prostaglandin E_2 (PGE_2) than those from normal mice. The addition of (+)-catechin at the beginning of incubation caused an inhibition of PGE_2 production of the adherent spleen cells. Further, the spleen cell suspension from CCl_4-treated mice gave a marked increase in generation of superoxide anions (O_2^-), which was also inhibited by the addition of (+)-catechin. From these results, it was found that the suppression of DTH was caused by adherent spleen cells in mice with experimental liver injury and eliminated by (+)-catechin owing principally to the inhibitory effects on the production of PGE_2 and O_2^- from the adherent spleen cells (Tajima *et al.*, 1985).

In studies of the effects of (+)-catechin on patients with acute viral hepatitis, there was a trend toward faster falls in serum bilirubin concentration and in transaminase activity in (+)-catechin-treated patients (Berengo and Esposito, 1975; Borel *et al.*, 1976; Seyfried *et al.*, 1975; Piazza *et al.*, 1983). Earlier elimination of HBsAg by (+)-catechin in patients with acute HB hepatitis has also been observed (Blum *et al.*, 1977; Di Nola, 1980). However, other trials failed to observe such effects on the serum transaminases or the elimination of HBsAg in blood of patients (Vido *et al.*, 1980; Schomerus *et al.*, 1984).

Clinical trials with (+)-catechin in chronic hepatitis have been reported. Significantly greater reductions in plasma transaminase and globulin levels in chronic persistent hepatitis with (+)-catechin treatment than with placebo were observed (Piazza *et al.*, 1981). In another study, "polyphasic" hepatitis, a slowly resolving and recurrently rebounding form of acute viral hepatitis, appeared to be more responsive to (+)-catechin than to placebo (Laverdant, 1981). Further, (+)-catechin plus adrenocorticosteroids gave greater improvement in patients with chronic hepatitis than placebo plus steroids, as judged by histological observations (Demeulenaere *et al.*, 1981).

Studies were performed in which (+)-catechin and placebo administration were compared for the prevention of hepatitis induced by the treatment of the antituberculosis drugs, isoniazid and rifampin (Siemon, 1981), or isoniazid, rifampin, and ethambutol (Knoblauch *et al.*, 1981). The incidence of drug-induced hepatotoxicity was prevented by prophylactic treatment with (+)-catechin in both studies.

(+)-Catechin has been used in the treatment of liver diseases for over 10 years, mainly in Europe. Although it may prove useful, the therapeutic efficacy of (+)-catechin is not quite clear. Clinically, (+)-catechin is known to have adverse actions such as the induction of haemolytic anaemia and fever (Neftel *et al.*, 1980; Brattig *et al.*, 1981). Further, patients with idiosyncrasy to (+)-catechin have died after treatment with this drug and, therefore, a question has risen on the safety of (+)-catechin.

V. *SCHIZANDRA* LIGNOIDS

The crude drug "gomishi" is prepared from the ripe fruits of Chinese magnolia vine, *Schizandra chinensis* Baillon (Schizandraceae), and has been utilized for antitussive and tonic purposes in Oriental medicine.

Since the initial isolation of lignans having the dibenzocyclooctane skeleton (schizandrin and deoxyschizandrin) from its seed oil (Kochetkov *et al.*, 1961), more than 30 lignans possessing the same carbon skeleton have been isolated and characterized (Chen *et al.*, 1976; Liu *et al.*, 1978; Ikeya *et al.*, 1982).

An alcoholic extract of *Schizandra* kernels reduced elevated GPT levels in mice treated with CCl_4 or thioacetamide, while a water extract of the kernels and an ethanol extract of the shells of the seeds were reported to be ineffective (Bao *et al.*, 1974). On the basis of these findings, seven lignans were isolated from *Schizandra* kernels (Chen *et al.*, 1976) and tested for antihepatotoxic activity. Most of

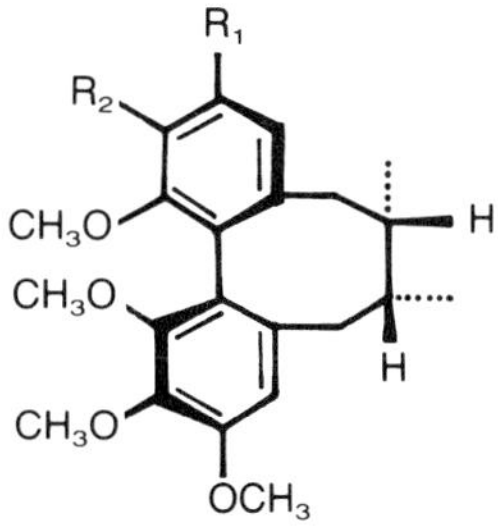

	R_1	R_2
Deoxyschizandrin	OCH_3	OCH_3
(±)-γ-Schizandrin	$-OCH_2O-$	

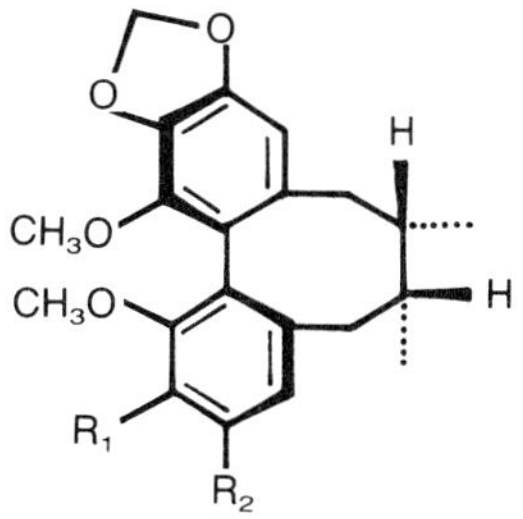

	R_1	R_2
Gomisin N	OCH_3	OCH_3
Wuweizisu C	$-OCH_2O-$	

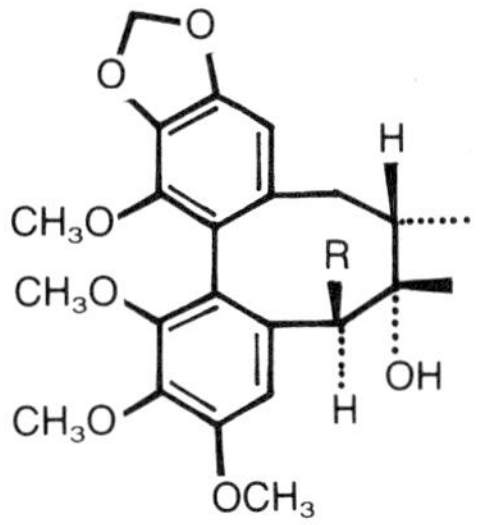

	R
Gomisin B	O-Ang
Gomisin C	O-Bz

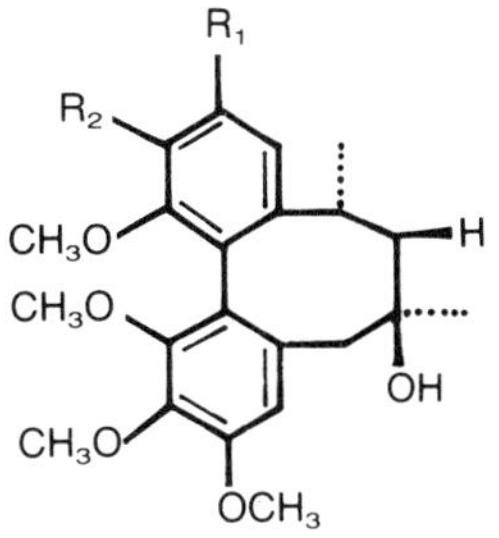

	R_1	R_2
Schizandrin	OCH_3	OCH_3
Gomisin A	$-OCH_2O-$	

Gomisin A, B and C, γ-schizandrin and wuweizisu C are given the following names by Chinese workers:

gomisin A=wuweizichin B=schizandrol B: gomisin B=schisantherin B=schisantherin A=wuweizi schizandrer B:gomisin C=schisantherin A=wuweizi ester A=schizandrer A:(+)-Gomisin K_3= schizantherol:wuweizisu C=schizandrin C: deoxyschizandrin=wuweizisu A=schizandrin A:(±) -γ-schizandrin=wuweizisu B=schizandrin B.

them prevented the elevation of serum GPT levels and morphological changes, such as inflammatory infiltration and liver cell necrosis induced by CCl_4 administration, although their potencies varied considerably. The intensity to lower the elevated GPT activity in the serum of mice treated with CCl_4 was found to follow the decreasing order gomisin B > gomisin A > wuweizisu C > γ-schizandrin > (+)-deoxyschizandrin > gomisin C ≥ deangeloylgomisin F. Similarly, the order of antihepatotoxic activity against thioacetamide-induced liver damage in mice was shown as gomisin B > gomisin A > wuweizisu C (Bao *et al.*, 1980). In fasting mice, the lignans stimulated glycogen synthesis, the order of the activity being gomisin A > (+)-deoxyschizandrin ≈ γ-schizandrin ≈ deangeloylgomisin F. The activity of gomisin A was comparable to that of cortisone. Since similar results were obtained in adrenalectomized mice, it was concluded that this effect is not dependent on the pituitary–adrenal system (Bao *et al.*, 1980).

Further, antihepatotoxic actions of *Schizandra* lignans were investigated using CCl_4-, GalN-, α-naphthylisothiocyanate-, and orotic-acid-induced hepatotoxicity in rats. The order of the intensity of inhibition was demonstrated as gomisin A ≫ dimethylgomisin J > (+)-deoxyschizandrin > schizandrin ≈ deoxygomisin A in CCl_4-induced liver damage, and gomisin A > schizandrin in GalN-induced lesions. Although gomisin A prevented orotic-acid-induced hepatotoxicity, prevention was not observed with gomisin A in α-naphthylisothiocyanate-induced liver injury. These findings suggest that the presence of the methylenedioxy grouping at C-12 and C-13 may contribute to the expression of antihepatotoxic activity (Maeda *et al.*, 1982). Further, marked prevention of wuweizisu C in liver injuries induced by CCl_4, GalN, and DL-ethionine were also found in rats by biochemical and morphological examinations (Takeda *et al.*, 1985).

Among gomisin A, schizandrin, deoxyschizandrin, deoxygomisin A, dimethylgomisin J, and wuweizisu C, which after i.p. injection displayed preventive effects on experimental liver lesions induced by various hepatotoxins, gomisin A was found to be most potent (Maeda *et al.*, 1982; Takeda *et al.*, 1985). Thus, the effects of oral administration of gomisin A were further examined. Gomisin A inhibited liver lesions produced by CCl_4, GalN and DL-ethionine, facilitated liver protein synthesis, elevated liver microsomal drug-metabolizing enzyme activities, and accelerated the proliferation of hepatocytes and the recovery of liver functions after partial hepatectomy in rats (Arai *et al.*, 1986).

The effects of gomisin A on liver functions in various experimental liver injuries and on bile secretion in CCl_4-induced liver injury have been studied. Although CCl_4, GalN and α-naphthylisothiocyanate markedly inhibited the disappearance of indocyanine green (ICG) from plasma, gomisin A showed a tendency to prevent the delay of excretion of plasma ICG induced by CCl_4, GalN, and orotic acid, but not that by α-naphthylisothiocyanate. Although bile flow, and biliary output of total bile acids and electrolytes (Na^+, K^+, Cl^-, and HCO_3^-) were decreased in CCl_4-treated rats, gomisin A maintained bile flow and biliary output of electrolytes nearly to the levels of the normal, but did not affect biliary output of total bile acids. These findings suggest that gomisin A possesses a liver-function-facilitating property in the liver of injured rats and that its preventive action of CCl_4-induced cholestasis is due to maintaining the function of the bile-acid-independent fraction (Maeda *et al.*, 1985).

Recently, a comprehensive assessment of the antihepatotoxic effects of 22 lignans from *Schizandra* fruit was carried out utilizing CCl_4- and GalN-induced cytotoxicity in primary cultured rat hepatocytes as model systems. Prominent protective actions were found with wuweizisu C and schisantherin D in CCl_4-produced cytotoxicity, and deoxygomisin A, gomisin N, wuweizisu C, gomisin C, and schisantherin D were effective in preventing GalN-induced cell damage. Examination of structure–activity relationships has confirmed that the methylenedioxy group of the dibenzocyclooctane skeleton may play an important role in antihepatotoxic activity (Hikino *et al.*, 1984a).

As previously described, CCl_4 is first metabolized to the CCl_3 radial, which results in lipid peroxidation, covalent binding to macromolecules, and inhibition of the calcium pump of microsomes, causing liver damage (Recknagel, 1983). Carbon tetrachloride is finally metabolized to carbon monoxide (CO), and NADPH and oxygen are consumed during CCl_4 metabolism in microsomes (Wolf *et al.*, 1980). Among the seven *Schizandra* lignans tested, (±)-γ-schizandrin, wuweizisu C, gomisin A, gomisin C, and gomisin B were shown to inhibit CCl_4-induced lipid perodixation and $[^{14}C]Cl_4$ covalent binding to lipids of liver microsomes from phenobarbital-treated mice. Those lignans also decreased CO production, and NADPH and oxygen consumption during CCl_4 metabolism by liver microsomes. These results suggest that the protective action of certain *Schizandra* lignans against CCl_4 hepatotoxicity may be due to the inhibitory effects of CCl_4-induced lipid peroxidation and to the binding of CCl_4-metabolites to liver microsomes (Liu and Lesca, 1982a). The mechanism of the inhibitory actions of wuweizisu C and gomisin A in CCl_4-induced liver damage has also been investigated by determining the effects of these lignans on lipid peroxidation. Although wuweizisu C and gomisin A exerted no inhibition on CCl_3 radical formation, both lignans inhibited CCl_4-, ADP/Fe^{3+}-, and ascorbate/Fe^{2+}-induced peroxidation, and wuweizisu C elicited stronger effects than gomisin A; this is in parallel with results on antihepatotoxic effects in CCl_4-induced cytotoxicity, indicating that the antioxidative action participates in the antihepatotoxic activity of wuweizisu C and gomisin A (Kiso *et al.*, 1985).

Further studies on the effect of *Schizandra* lignans on the liver monooxygenase system have been reported. Out of the eight lignans examined, (±)-γ-schizandrin, wuweizisu C, gomisin A, and schisanhenol induced increases of microsomal cytochrome P-450 concentration and NADPH cytochrome C reductase, aminopyrine demethylase and benzo(a)pyrene hydroxylase activities, as well as an increase in microsomal protein content. Remarkable enlargement of the smooth endoplasmic reticulum of liver cells was also observed. In order to characterize the types of cytochrome P-450 induced by these lignans, they were separated by a DEAE cellulose column and enzymatically and immunologically compared with the corresponding fractions obtained from phenobarbital-treated rats. Thus, a cytochrome P-450 species that showed the same chromatographic behaviour, possessed a common antigenic site with phenobarbital-induced cytochrome P-450, and exhibited a high benzphetamine demethylase activity was induced. Other cytochrome P-450 species were not affected. From these results, it was concluded that (±)-γ-schizandrin, wuweizisu C, gomisin A, and schisanhenol are phenobarbital-like inducers of microsomal monooxygenases (Liu *et al.*, 1981, 1982a; Liu and Lesca, 1982b; Liu, 1985).

In order to further clarify the biological consequence of the induction of cytochrome P-450 by *Schizandra* lignans, the effects of (±)-γ-schizandrin and gomisin A on benzo(*a*)pyrene metabolism and mutagenicity were investigated. The pattern of benzo(*a*)pyrene metabolites formed by catalysis of rat liver microsomes was modulated by (±)-γ-schizandrin and gomisin A treatments. Although benzo(*a*)pyrene-4,5-oxide and 9-hydroxybenzo(*a*)pyrene-4,5-oxide were increased markedly, benzo(*a*)pyrene-7,8-diol-9,10-epoxide, which is highly mutagenic and carcinogenic, was not increased. Pretreatment of rats with gomisin A and benzo(*a*)pyrene in combination, decreased the capacity of rat liver microsomes to activate benzo(*a*)pyrene to its mutagenic metabolites, as judged by the Ames test. The antimutagenic effect of dual induction by gomisin A and benzo(*a*)pyrene correlated with a decrease of covalent binding of benzo(*a*)pyrene metabolites to DNA. These results show that the induction of cytochrome P-450 by certain *Schizandra* lignans is not only harmless but it may be beneficial to the body (Liu and Lesca, 1982b).

More recently, the effect of gomisin A on experimental fulminant hepatitis was investigated using mice. Thus, acute fulminant hepatitis was induced by the intravenous injection of heat killed *propionibacterium* acres and the following intravenous injection of lipopolysaccharide 7 days later. Simultaneous treatment with gomisin A (i.p.) and lipopolysaccharide increased the survival rate from 7.5 to 80% and, remarkably, prevented liver cell necrosis histologically. Although glucagon–insulin therapy or steroid treatment has been clinically employed for fulminant hepatitis, a satisfactory remedy has not been established until now. Gomisin A is expected to show efficacy in the treatment of fulminant hepatitis (Mizoguchi *et al.*, 1985b).

Clinically, *Schizandra* fruit was tested very early for tonic, especially neurotonic purposes (Liu, 1975). Since 1972, it has been reported to be effective for liver diseases (Liu, 1979). In one such clinical trial, the total active principles of *Schizandra* fruit, extracted with ethanol, were administered to patients with chronic hepatitis. A total of 189 cases of chronic viral hepatitis with elevated serum GPT levels were selected for treatment with the alcoholic extract. Among them, 107 cases were given 60–100 mg of the extract (corresponding to 1.5 g of the crude material) and the remaining 82 cases received a liver extract–vitamin E complex as control. After 16–24 weeks of treatment, 73 out of 107 cases treated with the alcoholic extract of *Schizandra* fruit showed a fall of serum GPT to normal levels. No rebound was observed after withdrawal of the drug. The rate of effectiveness in lowering the GPT level was 68.2% in the treated group and 44% in the control group. Moreover, the average time needed for lowering the GPT level from elevated to normal was about 4 weeks for the treated group and 8 weeks for the control group. Differences in the effective rate and the average time needed were statistically significant between the two groups. However, improvement of other liver function tests was found to be less pronounced. An alcoholic extract of *Schizandra* fruit was also found to be more effective in relieving symptoms of sleeplessness, fatigue, abdominal tension, and loose bowels. No serious side-effects of an alcoholic extract of *Schizandra* fruits were observed except that 4 out of 107 cases developed mild and transient nausea, headache, and stomach-ache (Liu, 1977).

During the course of the synthesis of wuweizisu C, an active component isolated from *Schizandra* fruit, biphenyl dimethyl dicarboxylate (DDB); in fact (±)-α-DDB was prepared as an intermediate (Xie *et al.*, 1981). Biphenyl dimethyl dicarboxylate was found to inhibit the increase of GPT activity of mice treated with CCl_4 or thioacetamide, which was not associated with direct inhibitory action on GPT activity, as estimated by an *in vitro* experiment (Liu *et al.*, 1979). In addition, DDB decreased liver GPT activity of the normal, CCl_4-intoxicated, and prednisolone-treated mice, although no remarkable effects on GOT or aldolase activities were observed. Although the liver GPT was significantly decreased by DDB, the heart GPT still remained at normal levels. When the serum of mice given DDB was incubated with a liver homogenate of CCl_4-intoxicated mice *in vitro*, no decrease of GPT activity of liver homogenate was observed. Further, no acceleration of the spontaneous decrease by DDB of serum GPT levels in mice injected with a high GPT serum (from CCl_4-intoxicated mice) was found. Therefore, it seems that the GPT lowering action of DDB is neither due to direct inhibition of serum and liver GPT activities nor due to an increase of spontaneous disappearance of GPT from mouse blood circulation (Liu *et al.*, 1982b).

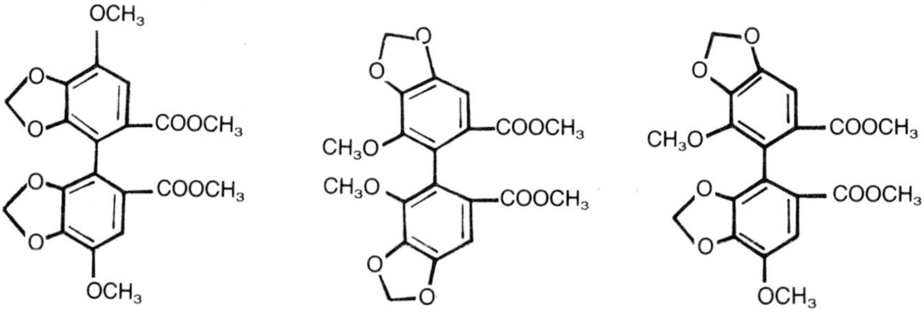

(±)-α-DDB (±)-β-DDB (±)-γ-DDB

In reality, three DDB congeners were prepared and a study of their structure–activity relationships demonstrated that the order of the intensity of the antihepatotoxic effect on CCl_4-induced liver damage was (±)-γ-DDB > (±)-α-DDB > (±)-β-DDB. Because of availability, α-DDB was subjected to clinical trials as described below. Contribution of the optical activity of the α-DDB congeners was quite interesting because the order of activity was shown to be (+)-α-DDB > (±)-α-DDB > (−)-α-DDB (G.-T. Liu, personal communication, 1986).

DDB was tested clinically for the treatment of viral hepatitis. Rates of recovery of elevated serum GPT levels to normal was demonstrated to be 72.8% at 2 weeks after treatment and 78.8% at 1 month; no side-effects were observed (Wang *et al.*, 1982).

VI. SAIKOSAPONINS

From ancient times, the crude drug "saiko", from the roots of *Bupleurum falcatum* L. or allied plants (Umbelliferae), has been widely used as a main component in prescriptions for treatment of feelings of fullness and oppression in the chest, or hypochondria mainly due to hepato-biliary diseases in Oriental medicine.

Recently, a series of triterpenoid saponins, saikosaponins a, b_1, b_2, b_3, b_4, c, d, e, and f, has been isolated from *Bupleurum* roots and their chemical structures have been determined. Later, it was revealed that among those saponins, saikosaponins b_1 and b_3, and saikosaponin b_2 were artifacts derived during processing from saikosaponin a and saikosaponin d, respectively (Kubota and Hinoh, 1968; Shimaoka *et al.*, 1975).

Saikosaponin a: R_1 = D-Glc-β(1→3)-D-Fuc
R_2 = CH_2OH R_3 = β-OH
Saikosaponin c: R_1 = D-Glc-β(1→6) / L-Rha-α(1→4) > D-Glc
R_2 = CH_3 R_3 = β-OH
Saikosaponin d: R_1 = D-Glc-β(1→3)-D-Fuc
R_2 = CH_2OH R_3 = α-OH

Saikosaponin b_1: R = β-OH
Saikosaponin b_2: R = α-OH

Concerning the pharmacological actions of saikosaponins, it has been reported that anti-inflammatory effects were observed for a mixture of saikosaponins a, b_1, b_2, c, and d, among which saikosaponins a and d showed the strongest anti-inflammatory activity (Takagi and Shibata, 1969; Yamamoto *et al.*, 1975b, c).

Remarkable increases of the serum GOT and GPT levels and histological deterioration of liver tissue caused by GalN administration to rats were inhibited by treatment with a saikosaponin fraction, mainly composed of saikosaponins b_1 and c, 2 hr after injection of GalN. Pretreatment with the saikosaponin fraction 2 hr before GalN injection was more effective for inhibition of hepatic injury induced by GalN (Arichi *et al.*, 1978). Later, it was revealed that in rats pretreated with each pure saikosaponin, especially saikosaponin a or d, remarkable inhibition of GalN-induced hepatic injury was observed (Abe *et al.*, 1980).

Investigation of the effects of saikosaponin d on acute hepatic injury or chronic hepatic injury produced by CCl_4 demonstrated that pretreatment with saikosaponin d mediated a remarkable inhibitory action on acute hepatic injury by CCl_4. A significant inhibition of lipid peroxidation induced by a single dose of CCl_4 in the liver of rats pretreated with saikosaponin d was also found. Although continuous injection of CCl_4 caused liver cirrhosis in rats, the severity of cirrhosis was reduced by simultaneous treatment with saikosaponin d (Abe, H. *et al.*, 1982).

A further study was carried out on the effect of saikosaponin d on CCl_4 hepatotoxicity enhanced by phenobarbitone and phenobarbitone-induced enzyme induction. Saikosaponin d showed protection aginst CCl_4 hepatotoxicity enhanced by phenobarbitone, and inhibition of increases in the cytochrome P-450 content and NADPH–cytochrome c reductase activity induced by phenobarbitone treatment, but the spectral characteristics of cytochrome P-450 were not altered by saikosaponin d treatment. Lipid peroxidation produced by NADP and CCl_4 was significantly lowered in the liver of rats pretreated with both phenobarbitone and saikosaponin d in comparison with that of rats pretreated with phenobarbitone alone (Abe *et al.*, 1985).

In order to clarify the mechanism of antihepatotoxic action of saikosaponins, the action of the saikosaponin fraction on intracellular microstructures of normal rat liver cells was first examined by electron microscopy. Thus, 24 hr after injection of the saikosaponin fraction at a dose of 2 mg/kg, the volume of endoplasmic reticulum, especially that of smooth endoplasmic reticulum, was doubled compared with controls, while that of mitochondria was unchanged. On the other hand, significant increases in the volume of mitochondria and the Golgi complex were observed in the liver of rats treated with the saikosaponin fraction (20 mg/kg), although no increase of the endoplasmic reticulum was found (Abe *et al.*, 1978a). The activities of glucose-6-phosphate and NADPH–cytochrome c reductase of the liver of rats treated with the saikosaponin fraction were compared with those of normal liver. Remarkable increases of both enzyme activities were observed in the liver of rats treated with the saikosaponin fraction at a dose of 2 mg/kg i.p., although no changes of those activities were found in rats injected with the saikosaponin fraction at a dose of 20 mg/kg i.p. (Abe *et al.*, 1978b).

Administration of saikosaponin a or d produced marked decreases of activities of microsomal enzymes, glucose-6-phosphatase, and NADPH–cytochrome c reductase, and a significant increase in activity of 5′-nucleotidase, whereas saikosaponins b_1, b_2, and c caused no such changes in these enzyme activities (Abe *et al.*, 1980).

Antihepatotoxic actions of the saikosaponins in CCl_4-induced liver injury are thus possibly due to decreases of microsomal enzyme activities and the inhibition of lipid peroxidation.

Further, it may be possible that the saikosaponins act on a common target (cell membranes) in the sequence of liver damage. Thus, although the saikosaponins were found to have a remarkable haemolytic potency at high concentrations (Abe *et al.*, 1978c), low concentrations of the saikosaponins protect rat erythrocytes from hypotonic- and heat-induced haemolysis. Modification of the aglycone part of the saikosaponins was found to cause marked changes on membrane stabilization. The saikogenins also protected erythrocytes from hypotonic haemolysis, but did not show any prevention of heat-induced haemolysis. It was thus suggested that the saikogenins react with erythrocyte membranes in a quite different manner from the parent saponins, and consequently combination with the sugar moiety plays an important role in the reaction with membranes. It is of interest to note that saikosaponins a and d, which have the lower transition concentrations from stabilization to lysis in the stabilization–lysis curves, possess stronger protective actions against liver injury than saikosaponin c, which shows the strongest potency in erythrocyte stabilization but has no protective action on hepatocytes (Abe *et al.*, 1981).

The effects of a water extract of *Bupleurum* root and the saikosaponins on immunologically induced cells with an extract of *Bupleurum* root or the saikosaponins (especially saikosaponin b_1)

prevented ADCC and activated macrophage-mediated cytotoxicity. Therefore, it was assumed that saikosaponins may exert protective effects on immunologically induced liver injury (Mizoguchi *et al.*, 1984b, 1985c). Further, the extract of *Bupleurum* root caused a significant increase in antibody production induced in human PBL after stimulation with PWM, which was attributable at least partially to the production of interleukin 1 from monocyte-macrophages, because a high interleukin 1 activity was detected in the culture supernatant of monocyte-macrophages treated with the extract of *Bupleurum* root. On the other hand, the extract of *Bupleurum* root exhibited a striking inhibitory effect on the transformation of human PBL, which was activated by mitogens such as PHA, PWM, or Con A (Mizoguchi *et al.*, 1985d). These results demonstrate that the biological actions of saikosaponins are quite similar to those of glycyrrhizin, as described previously.

Clinical trials for estimation of the effectiveness of saikosaponins on chronic hepatitis have been reported. Long-term p.o. administration of a mixture of saikosaponins a and b (3:2) at a very low dose of 6 mg/day showed remarkable reductions of serum GOT and GPT levels, and statistical significance was observed at 3, 6, and 12 months after the start of the medication with saikosaponins, compared with controls. (Yamamoto, M. *et al.*, 1975a, 1981).

The use of preparations containing *Bupleurum* root "sho-saiko-to" in traditional Oriental medicine has recently become popular with modern (Western) doctors of Japan for the treatment of chronic liver diseases. Thus, a number of clinical trials confirmed their effectiveness for hepatitis on scientific grounds. One such clinical trial was performed to evaluate the efficacy of long-term treatment with "sho-saiko-to" on patients with chronic hepatitis for 4 years. As a result it was found that significant improvements were observed in liver function, not only in the serum GOT, GPT, and γ-GTP levels but also on serum alkaline phosphatase levels and thymol turbidity tests after 2–3 months' medication. Improved conditions were maintained up to the end of the study (Yamamoto *et al.*, 1983, 1985). Even if *Bupleurum* preparations are effective for human hepatitis, it does not necessarily indicate that *Bupleurum* root itself really has hepatoprotective activity. However, because *Bupleurum* root is the main and common component of those preparations, it is quite probable that *Bupleurum* root contributes to their efficacy for liver diseases.

Although the accumulated data on the actions of the saikosaponins may not be enough to draw an unequivocal conclusion, it is likely from the results up to now that the saikosaponins may be helpful for the mitigation of liver diseases.

VII. DISCUSSION

As the most important representatives of natural products used for liver diseases, five drugs have been focused on in this chapter. Attention is drawn to the fact that these drugs are utilized chiefly on a regional basis. Thus, roughly speaking, the therapeutic effectiveness of silymarin and (+)-catechin is recognized in Europe, glycyrrhizin in Japan, *Schizandra* in China, and *Bupleurum* in Japan and China. Although one reason could be that the efficacy of the drugs varies depending on racial specificity, it may not be critical. Administrative policies may be partly responsible for limitation of their free utilization internationally. One major reason could be differences in tradition, popularity and/or availability of the original crude drugs in each region. The other reason of significance may be the differences in the main causes of liver disease from region to region (e.g., alcoholic intoxication is more prevalent in Europe than in Asia). Results of comparative studies on therapeutic effects in different regions should be interesting if those drugs become used worldwide.

For the evaluation of antihepatotoxic activity, hepatotoxin-induced liver damage has been utilized mainly so far becuase: (i) the assay procedures are simple; (ii) no other suitable assay methods having more related mechanisms to human liver disease are readily available; (iii) substances that reveal inhibitory activity in these assay methods exhibit therapeutic effectiveness for human hepatitis, more or less. The last finding may be rationalized provided that hepatotoxin-produced assay methods can assess a common target in the sequence of liver cell necrosis, e.g., protective activity on cell membranes. In the case of human hepatitis, when a hepatotoxin causing the liver disorder can be specified, the use of liver lesion models employing the specific hepatotoxin for evaluation of hepatoprotective activity is quite reasonable. However, in cases of viral hepatitis, pathological changes of liver tissue are essentially different from those produced by, for example, CCl_4, the most commonly utilized hepatotoxin. Therefore, doubt may be raised whether assay methods using hepatotoxin-induced liver damage models can adequately assess therapeutic effects for viral hepatitis. Although it was formerly considered that viral hepatitis was directly attributed to harmful effects of the virus itself, after the recognition of LSP as a liver membrane antigen, it has become clear that autoimmunity against this antigen participates in pathogenesis of chronic hepatitis evoked not only by the virus but also by ethyl alcohol and other chemicals.

Recently, *in vitro* models for immunologically induced liver damage have been introduced using primary cultured hepatocytes, and comparison of the antihepatotoxic effects of some liver-protective natural products employing these models with those assessed using hepatotoxin-produced lesion models has disclosed that they exhibit fairly different responses depending on the two types of assay (Mizoguchi *et al.*, 1981a, 1982; Feutren *et al.*, 1984; Shiki *et al.*, 1984; Kiso *et al.*, 1987a, b). On the basis of the above findings, it may thus be concluded that assays using immunologically induced liver damage models are more desirable for the evaluation of the efficacy of drugs on liver diseases, especially on viral hepatitis. Further, beside the *in vitro* acute liver cell lesions mentioned above, experimental autoimmune chronic hepatitis has been more recently devised in mice *in vivo* by repeated immunization with liver antigens including LSP for long terms (Mori *et al.*, 1984; Kiso *et al.*, unpublished). Therefore, more efficient development of effective drugs for liver disease may be accomplished when keeping in view the results obtained by *in vitro* and *in vivo* assay methods utilizing immunologically induced liver lesions that are more related to the causes responsible for human hepatitis.

REFERENCES

Abe, H., Konishi, H., Arichi, S., and Odashima, S. (1978a). *Kanzo* **19**, 1053.

Abe, H., Nagao, T., and Arichi, S. (1978b). Kanzo **19**, 1508.

Abe, H., Orita, M., Konishi, H., Arichi, S., and Odashima, S. (1985). *J. Pharm. Pharmacol.* **37**, 555.

Abe, H., Sakaguchi, M., Anno, M., and Arichi, S. (1981). *Naunyn-Schmiedeberg's Arch. Pharmakol.* **316**, 262.

Abe, H., Sakaguchi, M., Konishi, H., Tani, T., and Arichi, S. (1978c). *Planta Med.* **34**, 160.

Abe, H., Sakaguchi, M., Odashima, S., and Arichi, S. (1982). *Naunyn-Schmiedeberg's Arch. Pharmakol.* **320**, 266.

Abe, H., Sakaguchi, M., Yamada, M., Arichi, S., and Odashima, S. (1980). *Planta Med.* **40**, 366.

Abe, N., Ebina, T., and Ishida, N. (1982). *Microbiol. Immunol.* **26**, 535.

Alvisi, V., Bagni, B., D'Ambrosi, A., and Groothold, G. (1974). *Minerva Med.* **65**, 3175.

Arai, I., Takeda, S., Aburada, M., and Hosoya, E. (1986). *Abstr. Pap., 6th Symp. on the Development and Application of Naturally Occurring Drug Materials*, p. 92.

Arichi, S., Konishi, H., and Abe, H. (1978). *Kanzo* **19**, 430.

Bao, T.-T., Liu, G.-T., and Song, Z.-Y. (1974). *Chin. Med. J.* **54**, 275.

Bao, T.-T., Liu, G.-T., Song, Z.-Y., Xu, G.-F., and Sun, R.-H. (1980). *Chin. Med. J.* **93**, 41.

Berengo, A., and Esposito, R. (1975). *In* "New Trends in the Therapy of Liver Diseases" (A. Bertelli, ed.), p. 182. Karger, Basel.

Berenguer, J., and Carrasco, D. (1977). *Muench. Med. Wochenschr.* 240.

Bertelli, A. (1975). *In* "New Trends in the Therapy of Liver Diseases" (A. Bertelli, ed.), p. 92. Karger, Basel.

Blum, A. L., Doelle, W., Kortum, K., Peter, P., Strohmeyer, G., Berthet, P., Goebell, H., Pelloni, S., Poulsen, H., and Tygstrup, N. (1977). *Lancet* **2**, 1153.

Borel, G. A., Schelling, J. L., and Magnenat, P. (1976). *Z. Gastroenterol.* **1**, 24.

Braatz, R. (1976). *In* "Symposium on the Pharmacodynamics of Silymarin, Coloque Nov. 1974" (R. Braatz and C. C. Schneider, eds.), p. 44. Urban and Schwarzenberg, Munich, Berlin, Vienna.

Brattig, N., Diao, G. J., and Berg, P. A. (1981). *In* "International Workshop on (+)-Cyanidanol-3 in Diseases of the Liver" (H. O. Conn, and C. Wood, eds.), p. 227. Academic Press, London.

Chen, Y.-Y., Shu, Z.-B., and Li, L.-N. (1976). *Sci. Sin.* **19**, 276.

Cochrane, A. M. G., Moussouros, A., Thomson, A. D., Eddleston, A. L. W. F., and Williams, R. (1976). *Lancet* **1**, 441.

Demeulenaere, F., Desmet, V. J., and Dupont, E. (1981). *In* "International Workshop on (+)-Cyanidol-3 in Diseases of the Liver" (H. O. Conn and C. Wood, eds.), p. 135. Academic Press, London.

Desplaces, A. (1976). *In* "Symposium on the Pharmacodynamics of Silymarin, Coloque Nov. 1974" (R. Braatz and C. C. Schneider, eds.), p. 103. Urban and Schwarzenberg, Munich, Berlin, Vienna.

Desplaces, A., Choppin, J., Vogel, G., and Trost, W. (1975). *Arzneim.-Forsch.* **25**, 89.

Di Nola, F. (1980). *Lancet* **2**, 1379.

Djeu, J. Y., Heinbaugh, J. A., Holden, H. T., and Herberman, R. B. (1979). *J. Immunol.* **122**, 175.

Feutren, G., Lacour, B., and Bach, J. F. (1984). *J. Immunol. Methods* **75**, 85.

Fintelmann, V., and Albert, A. (1980). *Therapiewoche* **30**, 5589.

Fujisawa, K., Watanabe, Y., and Kimura, K. (1980). *Asian Med. J.* **23**, 745.

Fujisawa, K., Watanabe, Y., Kitahara, T., Kimura, K., Kawase, H., and Fukuzawa, K. (1984). *Shindan To Chir.* **72**, 40.

Gajdos, A., Gajdos-Torok, M., and Horn, R. (1970a). *C. R. Soc. Biol.* **164**, 1967.

Gajdos, A., Gajdos-Torok, M., and Horn, R. (1970b). *C. R. Soc. Biol.* **164**, 2187.

Gajdos, A., Gajdos-Torok, M., and Horn, R. (1972a). *Biochem. Pharmacol.* **21**, 594.

Gajdos, A. Gajdos-Torok, M. and Horn, R. (1972b). *C. R. Soc. Biol.* **166**, 277.

Gajdos, A., Gajdos-Torok, M., and Horn, R. (1975). *In* "New Trends in the Therapy of Liver Diseases" (A. Bertelli, ed.), p. 114. Karger, Basel.

Hahn, G., Lehmann, H. D., Kurten, M., Uebel, H., and Vogel, G. (1968). *Arzneim.-Forsch.* **18**, 698.

Handa, S. S., Sharma, A., and Chakraborti, K. K. (1986). *Fitoterapia* **57**, 307.

Hammerl, H., Pichler, O., and Studlar, M. (1971). *Med. Klin.* **66**, 1204.

Hennings, G. (1979). *Arzneim.-Forsch.* **29**, 720.

Hikino, H. (1985). *In* "Economic and Medicinal Plant Research, Vol. I" (N. Farnsworth, H. Hikino and H. Wagner, eds.), p. 53. Academic Press, London.

Hikino, H., Kiso, Y., Taguchi, H., and Ikeya, Y. (1984a). *Planta Med.* **50**, 213.

Hikino, H., Kiso, Y., Wagner, H., and Fiebig, M. (1984b). *Planta Med.* **50**, 248.

Hino, K., Miyahara, T., Miyagawa, H., Fujikura, M., Iwasaki, M., and Takahashi, J. (1981). *Kan Tan Sui* **3**, 137.

Holzgartner, H. (1970). *Therapiewoche* **20**, 1868.

Hopf, U., Meyer zum Buschenfelde, K. H., and Arnold, W. (1976). *N. Engl. J. Med.* **294**, 578.

Hopf, U., Meyer zum Buschenfelde, K. H., and Freundenberg, J. (1974). *Clin. Exp. Immunol.* **16**, 117.

Ikeda, Y., Kawakami, K., Sato, I., Tajima, S., Ito, K., and Nose, T. (1984a). *J. Pharm. Dyn.* **7**, 15.

Ikeda, Y., Kawakami, K., Sato, I., Tajima, S., Ito, K., and Nose, T. (1984b). *J. Pharm. Dyn.* **7**, 21.

Ikeya, Y., Taguchi, H., and Yosioka, I. (1982). *Chem. Pharm. Bull.* **30**, 3207, and references cited therein.

Jensen, D. M., McFarlane, I. G., Portmann, B. S., Eddleston, A. L. W. F., and Williams, R. (1978). *N. Eng. J. Med.* **299**, 1.

Kakumu, S., Arakawa, Y., Goji, H., Kashio, T., and Yata, K. (1979). *Gastroenterology* **76**, 665.

Kakumu, S., Murakami, H., and Kuriki, J. (1983). *Clin. Exp. Immunol.* **52**, 430.

Kakumu, S., Yata, K., and Kashio, Y. (1980). *Gastroenterology* **79**, 613.

Kappus, H., Koster-Albrecht, D., and Remmer, H. (1979). *Arch. Toxicol.* **2** (Suppl.), 321.

Kasai, Y., Sasaki, E., Sekiguchi, S., Kaneko, M., and Sakamoto, S. (1981). *Igaku No Ayumi* **116**, 50.

Kashio, T., Hotta, R., and Kakumu, S. (1981). *Clin. Exp. Immunol.* **44**, 459.

Keppler, D., Lesch, R., Reutter, W., and Decker, K. (1968). *Exp. Mol. Pathol.* **9**, 279.

Kiesewetter, E., Leodolter, I., and Thaler, H. (1977). *Leber Magen Darm* **7**, 318.

Kiso, Y., Kato, S., Kawakami, Y., and Hikino, H. (1987a). *Phytotherapy Res.* (in press).

Kiso, Y., Kawakami, Y., Kikuchi, K., and Hikino, H. (1987b). *Planta Med.* (in press).

Kiso, Y., Tohkin, M., and Hikino, H. (1983a). *Planta Med.* **49**, 222.

Kiso, Y., Tohkin, M., and Hikino, H. (1983b). *J. Nat. Prod.* **46**, 841.

Kiso, Y., Tohkin, M., Hikino, H., Hattori, M., Sakamoto, T., and Namba, T. (1984). *Planta Med.* **50**, 298.

Kiso, Y., Tohkin, M., Hikino, H., Ikeya, Y., and Taguchi, H. (1985). *Planta Med.* 331.

Knoblauch, M., Suter, F., and Schneider, B. (1981). *In* "International Workshop on (+)-Cyanidanol-3 in Diseases of the Liver" (H. O. Conn and C. Wood, eds.), p. 221. Academic Press, London.

Koch, H. (1980). *Pharm. Unserer Zeit* **9**, 33, 65.

Kochetkov, N. K., Khorlin, A., Chizhov, O. S., and Sheichenko, V. I. (1961). *Tetrahedron Lett.* **730**, and references cited therein.

Koster-Albrecht, D., Koster, U., Kappus, H., and Remmer, H. (1979). *Toxicol. Lett.* **3**, 363.

Kubota, T., and Hinoh, H. (1968). *Tetrahedron Lett.* 303.

Kumada, H., Ikeda, K., Katsuki, T., Yoshida, I., and Yoshiba, A. (1983). *Igaku To Yakugaku* **9**, 881.

Laverdant, C. (1981). *In* "International Workshop on (+)-Cyanidanol-3 in Diseases of the Liver" (H. O. Conn and C. Wood, eds.), p. 131. Academic Press, London.

Liu, C.-S., Fang, S.-D., Huang, M.-F., Kao, Y.-L., and Hsu, J.-S. (1978). *Sci. Sin.* **21**, 483.

Liu, G.-T. (1985). *In* "Advance in Chinese Medicinal Materials Research" (H. M. Chang, H. W. Yeung, W.-W. Tso and A. Koo, eds.), p. 257. World Scientific Publications, Singapore.

Liu, K.-T. (1977). *Abstr. Pap., Seminar on the Use of Medicinal Plants in Health Care*, p. 13.

Liu, K.-T., Cresteil, T., Columelli, S., and Lesca, P. (1982a). *Chem.–Biol. Interact.* **39**, 315.

Liu, K.-T., Cresteil, T., Le Provost, E., and Lesca, P. (1981). *Biochem. Biophys. Res. Commun.* **103**, 1131.

Liu, K.-T., and Lesca, P. (1982a). *Chem.-Biol. Interact.* **41**, 39.

Liu, K.-T., and Lesca, P. (1982b). *Chem.–Biol. Interact.* **39**, 301.

Liu, K.-T., Wang, G.-F., Wei, H.-L., Bao, T.-T., and Song, Z.-Y. (1979). *Yao Hsueh Hsueh Pao* **14**, 598.

Liu, K.-T., Wei, H.-L., and Song, Z.-Y. (1982b). *Yao Hsueh Hsueh Pao* **17**, 101.

Liu, S. (ed.) (1975). "Zhong-yao Yan-jii Wen-xian Zhai-yao (Summary of Literature on Chinese Medicine Research) 1820–1941". Scientific Publishing House, Beijing.

Liu, S. (ed.) (1979). "Zhong-yao Yan-jii Wen-xian Zhai-yao (Summary of Literature on Chinese Medicine Research) 1962–1974". Scientific Publishing House, Beijing.

Machicao, F., and Sonnenbichler, J. (1977). *Hoppe Seyler's Z. Physiol. Chem.* **358**, 141.

Maeda, S., Sudo, K., Miyamoto, Y., Takeda, S., Shinbo, M., Aburada, M., Ikeya, Y., Taguchi, H., and Harada, M. (1982). *Yakugaku Zasshi* **102**, 579.

Maeda, S., Takeda, S., Miyamoto, Y., Aburada, M., and Harada, M. (1985). *Jpn. J. Pharmacol.* **38**, 347.

Magliulo, E., Gagliardi, B., and Fiori, G. P. (1978). *Med. Klin.* **73**, 1060.

Manns, K., Meyer zum Buschenfelde, K. H., Hutteroth, T. H., and Hess, G. (1980). *Clin. Exp. Immunol.* **42**, 263.

McLean, A. E. M., and Nuttall, L. (1978). *Biochem. Pharmacol.* **27**, 425.

Meyer-Burg, J. (1972). *Klin. Wochenschr.* **50**, 1062.

Meyer zum Buschenfelde, K. H., and Hopf, U. (1974). *Br. J. Exp. Pathol.* **55**, 498.

Meyer zum Buschenfelde, K. H., Kossling, F. K., and Miescher, P. A. (1972). *Clin. Exp. Immunol.* **11**, 99.

Meyer zum Buschenfelde, K. H., and Miescher, P. A. (1972). *Clin. Exp. Immunol.* **10** 89.

Milosavljevic, Z., and Eckenberg, H. P. (1972). *Arztliche Praxis* **24**, 69.

Mizoguchi, Y., Ikemoto, Y., Arai, T., Yamamoto, S., and Morisawa, S. (1984a). *Arerugi* **33**, 328.

Mizoguchi, Y., Katoh, H., Tsutsui, H., Yamamoto, S., and Morisawa, S. (1985a). *Gastroenterol. Jpn.* **20**, 99.

Mizoguchi, Y., Monna, T., Yamamoto, S., and Morisawa, S. (1982). *Gastroenterol. Jpn.* **17**, 360.

Mizoguchi, Y., Sawai, H., Tsutsui, H., Ikemoto, Y., Arai, T., Miyajima, Y., Sakagami, Y., Higashimori, T., Monna, T., Yamamoto, S., and Morisawa, S. (1984b). *Kanzo* **25**, 40.

Mizoguchi, Y., Shiba, T., Ohnishi, F., Monna, T., Yamamoto, S., and Morisawa, S. (1981a). *Hepato-gastroenterology* **28**, 250.

Mizoguchi, Y., Shiba, T., Ohnishi, F., Monna, S., Yamamoto, S., Nakai, K., Otani, S., and Morisawa, S. (1981b). *Hepato-gastroenterology* **28**, 254.

Mizoguchi, Y., Tsutsui, H., Miyajima, K., Sakagami, Y., and Morisawa, S. (1985b). *Abstr. Pap., 21st Annu. Meet. Jpn. Soc. Hepatol.*

Mizoguchi, Y., Tsutsui, H., Miyajima, K., Yamamoto, S., Kitamura, M., Kodama, C., and Morisawa, S. (1985c). *J. Med. Pharm. Soc. Wakan-Yaku* **2**, 27.

Mizoguchi, Y., Tsutsui, H., Yamamoto, S., and Morisawa, S. (1985d). *J. Med. Pharm. Soc. Wakan-Yaku* **2**, 330.

Molhuysen, J. A., Gerbrandy, J., de Vries, L. A., de Jong, J. C., Lenstra, L. B., Turner, K. P., and Borst, J. C. (1950). *Lancet* **2**, 381.

Mori, Y., Mori, T., Yoshida, H., Ueda, S., Iesato, K., Wakashin, Y., Wakashin, M., and Okuda, K. (1984). *Clin. Exp. Immunol.* **57**, 85.

Neftel, K., Diem, P., Gerber, H., De Weck, A. L., and Stucki, P. (1980). *Schweiz. Med Wochenschr.* **110**, 380.

Nakamura, T., and Ichihara, A. (1981). *Minophagen Med. Rev.* **26**, 203.

Okita, K., Noda, K., Kondo, N., and Mizuta, M. (1975). *Kanzo* **16**, 620.

Perrissoud, D., Maignan, M. F., and Auderset, G. A. (1981). *In* "International Workshop on (+)-Cyanidanol-3 in Diseases of the Liver" (H. O. Conn and C. Wood, eds.), p. 21. Academic Press, London.

Perrissoud, D., and Weibel, I. (1980). *Naunyn-Schmiedeberg's Arch. Exp. Pathol. Pharmakol.* **312**, 285.

Piazza, M., De Mercato, R., and Guadagnino, V. (1981). *In* "International Workshop on (+)-Cyanidanol-3 in Diseases of the Liver" (H. O. Conn and C. Wood, eds.), p. 123. Academic Press, London.

Piazza, M., Guadagnino, V., Picciotto, L., De Mercato, R., Chirianni, A., Orlando, R., and Golden, G. (1983). *Hepatology* **3**, 45.

Platt, D., and Schnorr, B. (1971). *Arzneim.-Forsch.* **21**, 1206.

Pompei, R., Flore, O., Marccialis, M. A., Pani, A., and Loddo, B. (1979). *Nature (London)* **281**, 689.

Rauen, H. M., and Schriewer, H. (1971). *Arzneim.-Forsch.* **21**, 1194.

Rauen, H. M., and Schriewer, H. (1973). *Arzneim.-Forsch.* **23**, 148.

Recknagel, R. O. (1983). *Life Sci.* **33**, 401.

Reutter, W., Hassels, B., and Lesch, R. (1975). *In* "New Trends in the Therapy of Liver Diseases" (A. Bertelli, ed.), p. 121. Karger, Basel.

Ryle, R. R., Chakraborty, J., and Show, G. K. (1981). *In* "International Workshop on (+)-Cyanidanol-3 in Diseases of the Liver" (H. O. Conn and C. Wood, eds.), p. 185. Academic Press, London.

Saba, P., Galeone, F., Salvadorini, F., Guarguaglini, M., and Troyer, C. (1976). *Gazz. Med. Ital.* **135**, 236.

Salmi, H. A., and Sarna, S. (1982). *Scand. J. Gastroenterol.* **17**, 517.

Sato, I., Takebe, H., Tajima, S., Ikeda, Y., Ito, K., and Nose, T. (1983). *Nippon Yakurigaku Zasshi* **81**, 539.

Schmidt, L. (1971). *Therapiewoche* **21**, 35.

Schomerus, H., Wiedmann, K. H., Dolle, W., Peerenboon, H., Strohmeyer, G., Balzer, K., Goebell, H., Durr, H. K., Bode, C., Blum, A. L., Frosner, G., Gerlich, W., Berg, P. A., and Dietz, K. (1984). *Hepatology* **4**, 331.

Schopen, R. D., and Lange, O. K. (1970). *Med. Welt* 691.

Schriefers, K. H., and Dietz, D. (1969). *Therapiewoche* **19**, 1545.

Schriewer, H., Badde, R., Roth, G., and Rauen, H. M. (1973a). *Arzneim.-Forsch.* **23**, 157.

Schriewer, H., Badde, R., Roth, G., and Rauen, H. M. (1973b). *Arzneim.-Forsch.* **23**, 160.

Schriewer, H., Penin, L., Rahmede, D., Riese, B., Ruther, N., The, L. G., Gebauer, B., Abu Tair, M., and Reuen, H. M. (1973c). *Arzneim.-Forsch.* **23**, 149.

Schriewer, H., and Rauen, H. M. (1973). *Arzneim.-Forsch.* **23**, 159.

Schultz, R. M., Papamatheakis, J. D., and Chirigos, M. A. (1977). *Science* **197**, 674.

Sekiguchi, S., Sakamoto, S., Kaneko, M., Sasaki, E., and Kasai, Y. (1982). *Gendai Iryo* **14**, 341.

Seyfried, H., Brunner, H., and Grabner, G. (1975). *In* "New Trends in the Therapy of Liver Diseases" (A. Bertelli, ed.), p. 177. Karger, Basel.

Shiki, Y., Shirai, K., Saito, Y., Yoshida, S., Wakashin, M., and Kumagai, A. (1984). *J. Med. Pharm. Soc. Wakan-Yaku* **1**, 11.

Shimaoka, A., Seno, S., and Minato, H. (1975). *J. Chem. Soc. Perkin Trans.* **1**, 2043.

Siegal, F. P., and Siegal, M. (1977). *J. Immunol.* **118**, 642.

Siemon, G. (1981). *In* "International Workshop on (+)-Cyanidanol-3 in Diseases of the Liver" (H. O. Conn and C. Wood, eds.), p. 217. Academic Press, London.

Sipos, J., Gabor, V., Toth, Z., Bartok, K., and Ribiczey, P. (1980). *Int. J. Tiss. Reac.* **2**, 167.

Slater, T. F., and Eakins, M. N. (1975). *In* "New Trends in the Therapy of Liver Diseases" (A. Bertelli, ed.), p. 84. Karger, Basel.

Slater, T. F., and Scott, R. (1981). *In* "International Workshop on (+)-Cyanidanol-3 in Diseases of the Liver" (H. O. Conn and C. Wood, eds.), p. 33. Academic Press, London.

Sonnenbichler, J., Goldberg, M., Hane, L., Madubunyi, I., Vogl, S., and Zetl, I. (1986). *Biochem. Pharmacol.* **35**, 538.

Sonnenbichler, J., and Pohl, A. (1980). *Hoppe Seyler's Z. Physiol. Chem.* **361**, 1757.
Suzuki, H., Kikuchi, K., Tateda, A., Abe, S., Sato, Y., Hirasawa, T., Katayama, T., Yano, M., Koga, M., Nakada, T., Kitajima, H., Inoue, Y., Kubota, K., and Matsuo, S. (1980). *Clin. Eval.* **8**, 197.
Suzuki, H., Ohta, Y., Takino, T., Fujisawa, K., and Hirayama, C. (1983). *Asian Med. J.* **26**, 423.
Tajima, S., Nishimura, N., and Ito, K. (1985). *Immunology* **54**, 57.
Takagi, K., and Shibata, M. (1969). *Yakugaku Zasshi* **89**, 1367.
Takahashi, K., Shibata, S., Yano, S., Harada, M., Saito, H., Tamura, Y., and Kumagai, A. (1980). *Chem. Pharm. Bull.* **28**, 3449.
Takeda, S., Funo, S., Iizuka, A., Kase, Y., Arai, I., Ohkura, Y., Sudo, K., Kiuchi, N., Yoshida, C., Maeda, S., Aburada, M., and Hosoya, E. (1985). *Nippon Yakurigaku Zasshi* **85**, 193.
Valenzuela, A., Lagos, C., Schmidt, K., and Videla, L. A. (1985). *Biochem. Pharmacol.*, **34**, 2209.
Valloton, J. J., and Frei, P. C. (1981). *Infect. Immun.* **32**, 432.
Vazquez de Prada, J. R. (1974). *Med. Clin.* **62**, 28.
Videla, L. A., Fernandez, V., Valenzuela, A., and Ugarte, G. (1981). *Pharmacology* **22**, 343.
Vido, I., Schmidt, F. W., Muller, R., Ranft, U., Wildhirt, E., Holzer, E., Wallnofer, H., and Korb, G. (1980). *Dtsch. Med. Wachenschr.* **105**, 330.
Vogel, G. (1968). *Arzneim.-Forsch.* **18**, 1063.
Vogel, G., and Temme, I. (1969). *Arzneim.-Forsch.* **19**, 613.
Vogel, G., Trost, W., Braatz, R., Odenthal, K. P., Brusewitz, G., Antweiler, H., and Seeger, R. (1975). *Arzneim.-Forsch.* **25**, 82, 179.
Wagner, H. (1981). *In* "Natural Products as Medicinal Agents" (J. L. Beal and E. Reinhard, eds.), p. 217. Hippokrates, Stuttgart.
Wagner, H., Diesel, P., and Seitz, M. (1974). *Arzneim.-Forsch.* **24**, 466.
Wagner, H., Horhammer, L., and Munster, R. (1968). *Arzneim.-Forsch.* **18**, 688.
Wang, C.-F., Zhang, Y.-X., Gan, H.-Q., Shi, J.-Y., Fu, J.-H., Sun, F., Zhou, J.-Z., Ge, W.-J., Chen, S.-M., Zheng, W.-Y., Miao, Z.-Q., and Huang, R.-D. (1982). *Tian Jin Yi Yao* 93.
Watari, N. (1972). *J. Clin. Electron Microsc.* **5**, 81.
Watari, N. (1973). *Minophagen Med. Rev., Suppl.* **10**, 73.
Watari, N. (1975). *J. Clin. Electron Microsc.* **8**, 165.
Watari, N. (1977). *Minophagen Med. Rev.* **22**, 170.
Watari, N., and Torizawa, K. (1972). *J. Clin. Electron Microsc.* **5**, 199.
Wilhelm, H., and Haase, W. (1973). *Therapiewoche* **23**, 3276.
Wolf, C. R., Harrelson, W. G., Nastainczyk, W. M., Philpot, R. M., Kalyanaraman, B., and Mason, R. P. (1980). *Mol. Pharmacol.* **18**, 553.
Xie, J.-X., Zhou, J., Zhang, C.-Z., Yang, J.-H., Chen, J.-X., and Jin, H.-Q. (1981). *Yao Hsueh Hsueh Pao* **16**, 306.
Yamada, T., Ludwig, S., Kuhlenkamp, J., and Kaplowitz, N. (1981). *J. Clin. Invest.* **67**, 688.
Yamamoto, M., Hayashi, Y., Imadaya, A., Tosa, H., Hirayama, A., and Kumagai, A. (1975a). *Proc. Symp. Wakan-Yaku* **9**, 141.
Yamamoto, M., Kumagai, A., and Yamamura, Y. (1975b). *Arzneim.-Forsch.* **25**, 1021.
Yamamoto, M., Kumagai, A., and Yamamura, Y. (1975c). *Arzneim.-Forsch.* **25**, 1240.
Yamamoto, M., Uemura, T., Nakama, S., Uemiya, M., Hara, H., Hayashi, Y., Tosa, H., Imadaya, A., Masuda, Y., and Kumagai, A. (1981). *Proc. Symp. Wakan-Yaku* **14**, 56.
Yamamoto, M., Uemura, T., Nakama, S., Uemiya, M., Kasayama, S., Kishida, Y., Yamauchi, K., Komuta, K., and Kumagai, A. (1983). *Proc. Symp. Wakan-Yaku* **16**, 245.

Yamamoto, M., Uemura, T., Nakama, S., Uemiya, M., Tanaka, T., and Minamoto, S. (1985). *J. Med. Pharm. Soc. Wakan-Yaku* **2**, 386.

Yamamoto, S., Maekawa, Y., Imamura, M., and Kushima, T. (1958). *Rinsho Naika Shonika* **13**, 73.

Yamamoto, S., Suzuki, H., and Oda, T. (1981). *In* "International Workshop on (+)-Cyanidanol-3 in Diseases of the Liver" (H. O. Conn and C. Wood, eds.), p. 143. Academic Press, London.

4

Potential Fertility-regulating Agents from Plants

AUDREY S. BINGEL
HARRY H. S. FONG
Program for Collaborative Research in the Pharmaceutical Sciences
College of Pharmacy, Health Sciences Center, University of Illinois at Chicago
Chicago, Illinois, U.S.A.

I. INTRODUCTION

The ideal contraceptive, universally acceptable to, and usable by, everyone, has not yet been discovered, and probably never will be. While technologically sophisticated methods such as oral contraceptives and intrauterine devices (IUDs) are in use in developing countries, their cost can be prohibitive; furthermore, access to them by rural populations may be inadequate, as may be rural patient follow-up by physicians to prevent/treat possible adverse reactions to their use. Even in the U.S.A., a developed country, the availability of effective, reversible means of birth control for women who cannot use oral contraceptives, e.g., because of age, smoking habits, pre-existing health contraindications, etc., has become a problem since the manufacturers of three (Lippes Loop, Cu-7, Copper T) of the four commonly used IUDs have withdrawn their products from the market, leaving only the Progestasert available,

ECONOMIC AND MEDICINE PLANT RESEARCH VOLUME 2
ISBN 0-12-730063-5

an IUD that needs to be replaced yearly. While more women (and men) may thus choose to become sterilized, increased non-use of contraception and increased use of less-effective contraceptive methods, are also possible. The use of high-dose estrogen (or of high-dose estrogen + progestagen) may be able to prevent implantation, but rather precise timing is necessary for effectiveness. Abortifacient agents such as the prostaglandins, on the other hand, are primarily used for terminating pregnancies of 12–20 weeks' duration. Thus, efforts to find additional/alternative methods of birth control, for primary use and/or as a back-up method, are as necessary as ever, if not more so. Of particular interest would be methods that might (a) interfere with the process of implantation, but not necessarily have to be given prior to the occurrence of implantation, (b) be effective if given within a couple of weeks after one's first missed menstrual period, and/or (c) be usable by, and acceptable to, the male.

This review will summarize recent experimental and clinical studies involving extracts and compounds from higher plants that may help to fill the continuing need for an adequately diverse armamentarium of first-line and back-up methods of birth control. Included among recent reviews concerning this subject are those by Bingel and Farnsworth (1980), Farnsworth *et al.* (1983), Fong (1984), Waller *et al.* (1985), and Kong *et al.* (1986b). The reader may consult these reviews, and references therein, for additional background information concerning the plants, and constituents thereof, to be discussed in the present review.

II. PROTEINACEOUS COMPOUNDS

α-Trichosanthin [so named to distinguish it from β-trichosanthin, a glycoprotein isolated from *Trichosanthes cucumeroides* Maxim. (Tam *et al.*, 1985; Yeung *et al.*, 1985)] is a polypeptide isolated from *Trichosanthes kirilowii* Maxim. (Cucurbitaceae) roots. It has been administered in China intra-amniotically to induce second-trimester abortion, and intramuscularly to promote uterine evacuation in cases of dead fetus *in utero*, missed abortion, and hydatidiform mole (Jin, 1985). The incidence of complete expulsion of the uterine contents ranged from 95/100 (second-trimester abortion) to 29/37 (hydatidiform mole); modal time between drug administration and occurrence of either complete or incomplete expulsion was 4–7 days for second-trimester abortion and 1–3 days for the other three conditions.

α-Trichosanthin has also been administered, intramuscularly or extra-amniotically, to induce first-trimester abortion (Jin *et al.*, 1981; Liu *et al.*, 1985), but only in conjunction with other medications: testosterone propionate i.m. plus reserpine i.m. in the latter study; and a decoction prepared from seven traditional Chinese medicinal plants p.o., plus 15-methyl-$PGF_{2\alpha}$ intravaginally, with or without testosterone propionate i.m., in the former. Success rates approximated 90%, with the modal administration to abortion interval approximating 3–7 days. The need for, and rationale behind, such combination therapies do not seem to have been discussed at length in the literature. Zhou *et al.* (1982), however, noted that the "abortion time was too long" when radix trichosanthis was used alone to terminate early pregnancy, and that incomplete placental expulsion and prolonged vaginal bleeding were common when 15-methyl-$PGF_{2\alpha}$ was used alone in such cases; reportedly, the overall success rate, including that of complete placental expulsion, was improved, and the duration of vaginal bleeding decreased, when the two drugs were used in combination.

The effects of α-trichosanthin, given at various times during pregnancy, have also been investigated in several laboratory species, and some comparisons have been made. Single i.p. doses of 2 mg of α-trichosanthin/rabbit were ineffective in interrupting pregnancy if given on Days 4, 6, 10, or 12, but were effective if given on Days 15, 17, or 22 (Chang *et al.*, 1979; Lau *et al.*, 1980). In the rat, 1-mg doses injected s.c. or i.p. on Day 10 were ineffective (Chang *et al.*, 1979), while such a dose given i.m. on Day 7 reportedly was effective (Zhou *et al.*, 1982). α-Trichosanthin (4 mg s.c. on Day 8) was ineffective in terminating pregnancy in the hamster (Chang *et al.*, 1979), and also was toxic. In the mouse, a single dose of 100 mcg (4 mg/kg) i.p. on Day 4 was reported to be ineffective (Chang *et al.*, 1979), and a dose of 200 mcg (8 mg/kg) i.p. on Day 3 was reported to be toxic. Zhou *et al.* (1982), on the other hand, found that a dose of 10 mg/kg injected s.c. on both Days 4 and 5 completely inhibited pregnancy; the results of additional experiments carried out by the latter workers, however, suggested that this dose of α-trichosanthin might cause weight loss. When Law *et al.* (1983) subsequently discussed their own results (*vide infra*), which agreed with those of Zhou *et al.* (1982) but not with those of Chang *et al.* (1979), they speculated that there may have been differences in purity/potency of the various α-trichosanthin preparations that had been used.

The effects of α-trichosanthin, and/or those of α- and/or β-momorcharin, two glycoproteins isolated from *Momordica charantia* L. (Cucurbitaceae) seeds (Tam *et al.*, 1985), on pre- and peri-implantation mouse embryos have also been investigated (Tam *et al.*, 1985; Law *et al.*, 1983; Chan *et al.*, 1984; Tam *et al.*, 1984b). Further *in vitro* development was impaired, depending to some extent on dosage, duration of exposure, and/or specific stage of the embryo at the time of exposure; furthermore, when such drug-exposed embryos were transferred to pseudopregnant recipients, their success at implanting and developing further, likewise, were impaired (Tam *et al.*, 1985; Tam *et al.*, 1984a). Additional experiments involved treating pseudopregnant mice with one or other of these three compounds. All three, for example, impaired the decidual reaction of the uterus of pseudopregnant mice to a mechanical stimulus (Tam *et al.*, 1985). α-Momorcharin (the only compound so tested), given intraperitoneally to pseudopregnant mice, impaired the implantation and further development of untreated transferred embryos, and it further impaired such for α-momorcharin-treated transferred embryos (Tam *et al.*, 1984a).

These same workers have also attempted to interrupt pregnancy in the mouse with one or other of these compounds. Unfortunately, in their experiments with α- and β-momorcharin administered on Day 12 (Yeung *et al.*, 1985), "aborted mice" were defined as those in which the number of dead fetuses was greater than 50% of the total number of implantation sites; apparently, therefore, some "aborted mice" must have been carrying at least one viable fetus, i.e., were pregnant! In the experiments involving the administration of α-trichosanthin, or of α- or β-momorcharin, on Days 4, 6, or 8 (Law *et al.*, 1983; Chan *et al.*, 1984), the responses apparently were all-or-none, i.e., abortion or survival of each entire litter. In a further study (Tam *et al.*, 1984b), "non-pregnant" mice were defined as *either* those totally lacking implantations *or* those in which all implantations were degenerated. Superficially, at least, results of these studies suggest that for interrupting pregnancy in the mouse, α-trichosanthin may be reasonably effective if given on Days 4 or 8, α-momorcharin if given on Days 1, 2 or 3, and α- and β-momorcharin if given on Days 4, 6, or 12 (Law *et al.*, 1983; Chan *et al.*, 1984; Tam *et al.*, 1984b). The future of the momorcharins as clinical abortifacients, however, at least with respect to their use early in the first trimester, is unclear at present. Chan *et al.* (1986) have shown recently that a 24-hr *in vitro* exposure to α- or β-momorcharin (50 or 100 mcg/ml) was lethal to varying percentages of 9- and 10-day cultured mouse embryos, and teratogenic to comparable or greater percentages of those embryos.

Pinellin is a globular protein isolated from the juice of *Pinellia ternata* (Thunb.) Breitenb. (Araceae) tuber; results of studies investigating its effects on fertility in the mouse were reviewed recently by Kong *et al.* (1986b). Apparently, it was effective in terminating pregnancy if given on Days 6 or 7, but not if given at other times; simultaneous administration of progesterone or chorionic gonadotropin reportedly antagonized pinellin's antifertility effects.

Pinellin's effects have also been studied in the rabbit (Chen *et al.*, 1984). Direct administration of 500 or 800 mcg of pinellin into the uterine horns of rabbits, five days post-mating, prevented implantation; lower intrauterine doses, and 8.5 mg/kg doses administered i.m. on Days 5–7 were ineffective. Pinellin also was injected into the uterine horns of 4-day pseudopregnant rabbits, and into those of rabbits four days post-fertile mating. Untreated blastocysts were transferred to the treated pseudopregnant rabbits, and the drug-exposed blastocysts from the treated rabbits were transferred to untreated pseudopregnant rabbits. Untreated blastocysts did not implant when transferred into pinellin-treated horns, nor did blastocysts implant in untreated horns after they had been allowed to remain in pinellin-treated horns for 18 hr; after having been allowed to remain in treated horns for only 3.5 hr, however, 6/18 blastocysts did implant following transfer. Despite the use of somewhat different experimental techniques, these results with pinellin are not unlike some of those discussed above for α-trichosanthin and the momorcharins (Tam *et al.*, 1985; Tam *et al.*, 1984a).

With respect to these four compounds, α-trichosanthin, α- and β-momorcharin, and pinellin, it appears that the problem of immunogenicity has been discussed only for the first (Kong *et al.*, 1986b; Tam *et al.*, 1984a). Intradermal hypersensitivity tests with α-trichosanthin, followed by intramuscular challenge with the drug, are in fact carried out before this agent is administered for the various therapeutic purposes discussed above (Jin, 1985; Liu *et al.*, 1985). That a woman would be able to have one or more subsequent pregnancies safely terminated, or other repeated trophoblastic disorders safely treated following previous treatment with α-trichosanthin, seems unlikely. Zhong and Wang (1983) performed skin sensitivity and serum antibody tests in 103 patients, 1–14 years after they had had a second-trimester abortion induced by α-trichosanthin. Approximately one-third of the 94 patients studied after 1–9 years had positive results in one or both tests; both tests were negative in the nine patients studied after 10–14 years. Among the 103 pregnant women who reportedly had never received α-trichosanthin, there were two positive skin-test results and nine positive serum-antibody test results. Since pinellin is a protein and the momorcharins are glycoproteins, it seems possible that immunogenicity could be a problem with these compounds as well.

III. YUAN-HUA

Four diterpenoid orthoesters, yuanhuacine (**1**), yuanhuadine (**2**), yuanhuafine (**3**), and yuanhuatine (**4**), have been isolated from the roots and/or flowers of *Daphne genkwa* Sieb. & Zucc. (Thymelaeaceae), and have been reported to be abortifacient in the human and/or monkey (Kong *et al.*, 1986b; Lin *et al.*, 1981). These compounds apparently are non-immunogenic (Kong *et al.*, 1986b); however, yuanhuafine has been reported to cause severe skin irritation (Wang *et al.*, 1982), while yuanhuacine and yuanhuadine emulsions, at 1 mg/ml concentrations, caused severe skeletal muscle irritation when

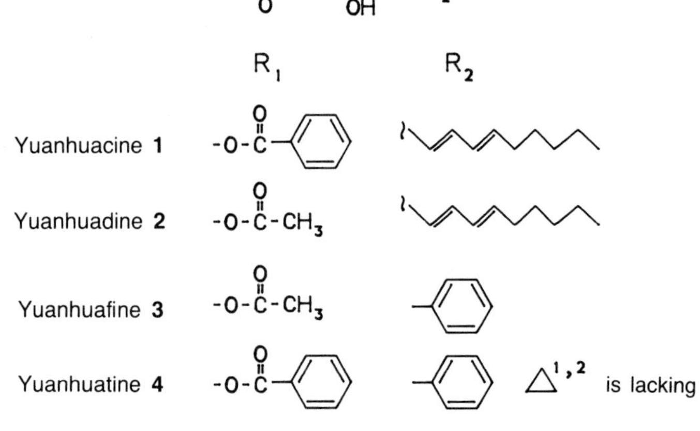

	R_1	R_2	
Yuanhuacine **1**	-O-C(=O)-C$_6$H$_5$		
Yuanhuadine **2**	-O-C(=O)-CH_3		
Yuanhuafine **3**	-O-C(=O)-CH_3		
Yuanhuatine **4**	-O-C(=O)-C$_6$H$_5$		$\Delta^{1,2}$ is lacking

given i.m. to rabbits (Lin *et al.*, 1981). Concern also has been expressed as to whether or not these compounds might be co-carcinogenic (Farnsworth *et al.*, 1983; Lin *et al.*, 1981), since some structurally related compounds such as simplexin and huratoxin do have this property; Lin *et al.* (1981) stated that experiments aimed to answer this question were in progress, but no results appear to have been published to date.

Clinically, yuanhuadine is administered intra-amniotically (Wang *et al.*, 1983), and yuanhuacine either intra- or extra-amniotically (Kong *et al.*, 1986b), to induce second-trimester abortion; the intra-amniotic route also was employed in a pre-clinical study of the abortifacient effect of these compounds in monkeys (Lin *et al.*, 1981). In the latter study, four monkeys receiving yuanhuadine doses of 50–5000 mcg, respectively, aborted approximately 1 day later, while five monkeys receiving yuanhuacine doses of 200–8000 mcg, respectively, aborted in approximately 1.5–2.5 days; lower doses of each compound, respectively, were ineffective. In a clinical study (Wang *et al.*, 1983) in which 60-mcg doses of yuanhuadine were employed, the administration to abortion interval reportedly averaged about 2 days.

In the monkey, at least, yuanhuadine thus appears to be more potent than is yuanhuacine. Even in the human, yuanhuacine reportedly is given at slightly higher doses (70–80 mcg/woman) than is yuanhuadine (60–70 mcg/woman) (Kong *et al.*, 1986b); furthermore, while the active dose of yuanhuafine has been reported to be 200 mcg in the monkey, that for yuanhuatine reportedly is only 50 mcg. This may help to explain the results of a clinical study reported by Huang (1982). Alcoholic extracts of yuan-hua flowers and of yuan-hua roots, and the compound yuanhuacine, respectively, were administered intra-amniotically to 427/442, 78/84, and 64/79 women who were 14–28 weeks pregnant; the remaining women received the respective preparations extra-amniotically. Success rate was high, approaching 100% with the extracts and 95% with yuanhuacine; the average administration to abortion interval for the extracts, however, was only 28 hr, as opposed to 40 hr for yuanhuacine. Notwithstanding the content of yuanhuacine being the highest (Wang *et al.*, 1986), the other three orthoesters have also been reported to be present in the root, at least, of *D. genkwa*. The effects of the root elixir, therefore, are likely to have been due to the combined actions of all four compounds, including two that may be more potent than is yuanhuacine, albeit present at a lower concentration.

Additional observations made in this study (Huang, 1982) argue both for and against the use of yuan-hua preparations (including yuanhuacine, itself) for inducing second- (and early-third-) trimester abortions. Fetal death had occurred by the time of abortion in almost 100% of the cases induced with yuan-hua, as opposed to absence of life signs in only 75/167 fetuses aborted following the use of radix trichosanthis. The average administration to abortion interval with the latter preparation, furthermore, was reported in this study to be 69 hr following intra-amniotic administration and 142 hr following i.m. On the other hand, use of the yuan-hua preparations (including yuanhuacine, alone) resulted, in about 3.5% of the cases, in very strong uterine contractions, associated in a few cases (flower extract, 3/442; root extract, 1/84) with rupture of the uterine neck. The administration of radix trichosanthis, in contrast, reportedly induced mild uterine contractions, perhaps explaining the longer latency to the occurrence of abortion; use of the latter preparation was not associated with uterine damage.

IV. ZOAPATLE

Zoapatle [*Cihuapahtli: cihua* = woman; *pahtli* = medicine (Senties and Amayo, 1964)] is the common name for *Montanoa tomentosa* (Cerv.) ssp. *tomentosa* (Funk) (Compositae) (Gallegos, 1983); it reportedly has been used in Mexico, for more than 500 years, in the form of a crude aqueous extract obtained by boiling its leaves in water for 10–20 min. According to classical books concerning indigenous remedies, zoapatle's use apparently was always in obstetrics, i.e., for inducing labor, for increasing the tone and frequency of uterine contractions during labor, and for decreasing *post-partum* bleeding (Gallegos, 1983); in 1866, its potential use as a menses inducer reportedly was first described. Its potential use as a fertility-regulating agent might also be implied from the report by Senties and Amayo (1964) that maternal or fetal accidents had been attributed at times to zoapatle, when the latter was administered to pregnant women before delivery.

The term, zoapatle, however, can actually refer to several *Montanoa* species (Estrada *et al.*, 1983), and mixtures of these species may be sold under the name, zoapatle, in Mexican markets. Thus, even though the intensive work carried out on *M. tomentosa* since the early 1970s (see Bingel and Farnsworth, 1980; Farnsworth *et al.*, 1983;

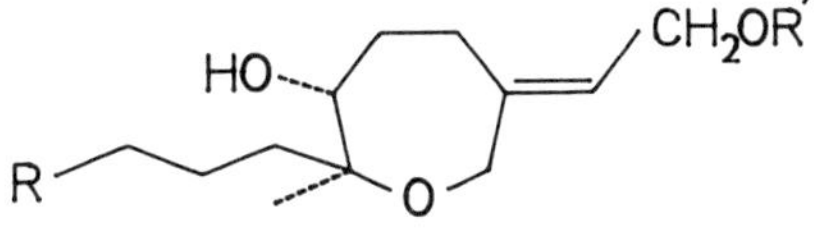

HO, CH_2OR', R, O

	R	R′
Zoapatanol **5**	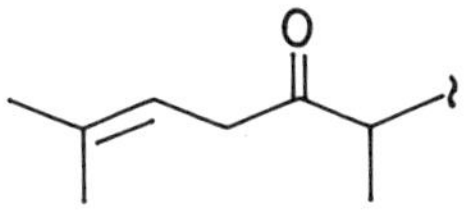	H
Montanol **6**	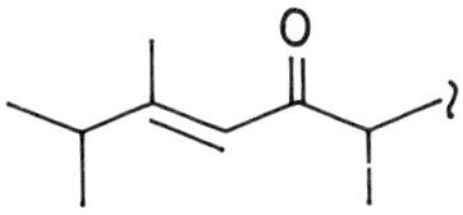	H
"9"	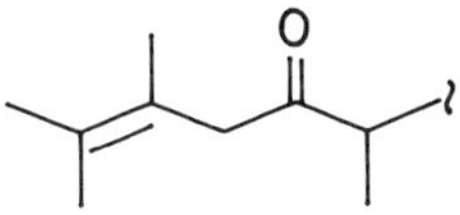	H
Tomentol **10**	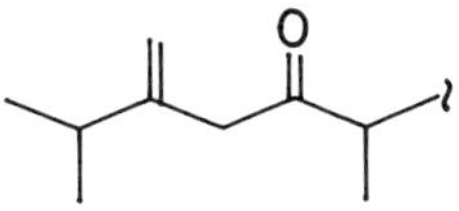	H
Tomexanthol **12**	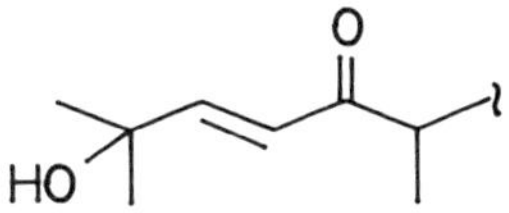 HO	H
21-Normontanol **13**	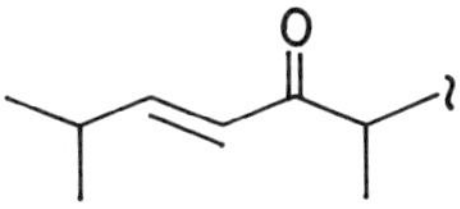	H
Tomexanthin **16**	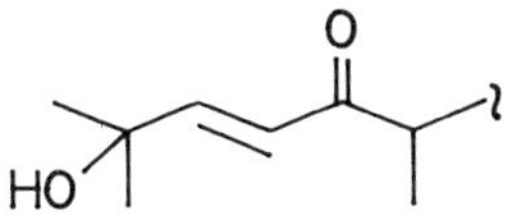 HO	Ac
Tomentanol **17**	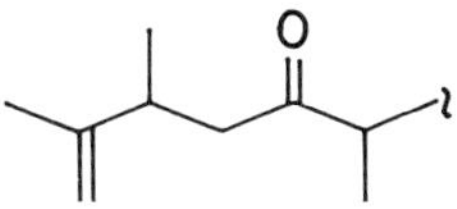	H

and Gallegos, 1983; and references therein) led to the isolation of the novel oxepane diterpenoids, zoapatanol (**5**) and montanol (**6**) [both of which reportedly caused uterine contractions in rabbits *in situ* (albeit montanol's requiring a 10-fold higher dose), zoapatanol also reportedly interrupting pregnancy in the guinea-pig], it is not surprising that several other *Montanoa* species have also recently been investigated (Fong, 1984; Gallegos, 1983; Estrada *et al.*, 1983) as possibly alternative/additional sources of zoapatanol and/or as sources of additional, potentially useful fertility-regulating agents.

Research concerning zoapatle has, in fact, proceeded along several different, but not necessarily mutually exclusive, pathways: (a) basic and clinical pharmacological testing of crude or semipurified *Montanoa* sp. extracts, with or without reported attempts to isolate the active component(s); (b) phytochemical studies of *Montanoa* species, with or without reported pharmacological testing of the isolated compounds; and (c) synthesis of zoapatanol and its analogs, usually with reports of pharmacological testing thereof. The following is a discussion of the results of such studies carried out in other laboratories as well as in our own.

Clinical studies with zoapatle have been few in number and have previously been briefly reviewed (Bingel and Farnsworth, 1980; Farnsworth *et al.*, 1983; Fong, 1984; Gallegos, 1983). The most recent of these studies (Landgren *et al.*, 1979) was small but controlled. Three of six 6–7-week (LMP) pregnant women received, over a 24-hr period, the equivalent of 124 g of dried leaves in the form of a tea (1 g/ml, following concentration under reduced pressure), while the other three received the equivalent of 160 g; six 6–8-week pregnant controls received corresponding volumes of a commercial tea. Although significant cervical dilatation and menstrual-like cramps occurred in all zoapatle-treated patients, abortion had not occurred by 24 hr after the last dose, at which time all pregnancies were terminated by vacuum aspiration. The doses used were limited by the Swedish Drug Regulatory Agency, and the possibility that higher doses might have been effective cannot be ruled out; furthermore, the observation period was rather short. The tea that was used in this study had been prepared by first grinding the dried leaves into a fine powder and then boiling the latter with 10 volumes of water. Such a total volume, prior to its being concentrated in the present study, does appear to compare favorably with that reportedly used by Aztec Indian women in the 1550s (Levine *et al.*, 1981); the latter were described as having ingested liters of zoapatle tea as a labor inducer and as an antifertility agent, but unfortunately the amount of leaves consumed was not stated.

Basic research papers concerning zoapatle are more numerous. A number concern the *in vitro* effects of zoapatle extracts, and/or isolated compounds, on contraction of uterine tissue from various species. Others concern the *in vivo* effects of such materials administered to various species, by various routes of administration, at various times *post-coitum*. Recently, in our laboratory, we have been attempting to correlate the *in vitro* and *in vivo* effects of such materials, using the estrogenized guinea-pig uterine strip and 22-day pregnant guinea-pig bioassays, respectively.

Presumably because of zoapatle's ethnomedical usage around the time of parturition (Gallegos, 1983), *in vitro* studies of this material and its constituents have been concerned largely with examining and comparing their uterine stimulant activity. Estrada *et al.* (1983), for example, using estrogenized guinea-pig uterine strips *in vitro*, reported the relative uterotonic potencies (expressed as mIU–/ml equivalents of oxytocin) of seven different *Montanoa* species cultivated in experimental plots and/or growing wild in various parts of Mexico. Ponce-Monter *et al.* (1983) further examined the effects of *Montanoa* sp. crude aqueous dialysed extract on estrogenized uterine strips from several species. Based on the published tracings, the extracts appeared to have had a stimulating action on guinea-pig and cat tissue, and perhaps also on that of the hamster, apparently no effect on that of the monkey, and a slightly inhibitory effect (decreased peak height and decreased frequency of contraction) on that of the rat. Lozoya *et al.* (1983) showed further tracings indicating that the contraction-stimulating effect of *M. tomentosa* aqueous crude extract on pregnant guinea-pig uterine strips *in vitro* could also be produced by a hexane extract of that plant, as well as by kauradienoic acid (**7**), a compound isolated from the latter extract. Gallegos (1985) has also reported on the *in vitro* effects of kauradienoic acid on

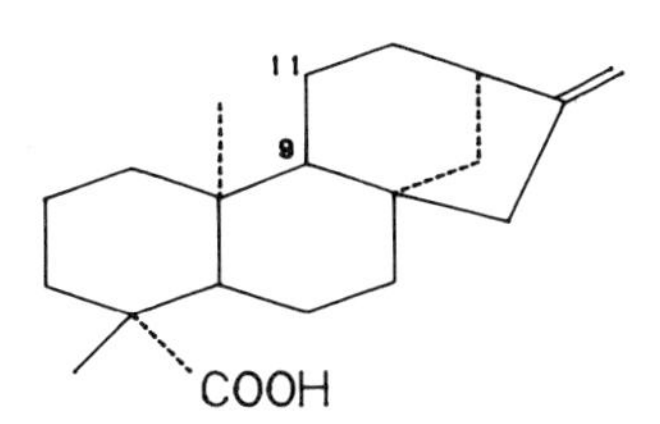

	Modification
Kauradienoic acid **7**	$\Delta^{9,11}$
Kaurenoic acid **8**	—

estrogenized guinea-pig uterine strips, comparing the effects of this compound with those of kaurenoic acid (**8**), a mixture of the latter two, zoapatanol (synthetic), and montanol. Unfortunately, only relative uterotonic potencies (kaurenoic acid, 2.60; kauradienoic acid, 2.15; mixture, 6.76; zoapatanol, 1.00; and montanol, 1.00) were presented; i.e., no tracings or absolute values were shown. Interestingly, the same synthetic (±)-zoapatanol was shown by Smith *et al.* (1981) to have had no effect on non-pregnant rabbit and rat uterine tissue *in vitro*.

Several groups of investigators have attempted to prevent/interrupt pregnancy in various species, with zoapatle extracts or constituents thereof. Successful results (absence of implants or substantial numbers of non-viable fetuses) are shown in Table I. In discussing the results of Pedrón *et al.* (1985), Gallegos (1985) speculated that the greater efficacy (total absence of implants) of intrauterine *M. frutescens*, in comparison with intrauterine *M. tomentosa* ssp. *tomentosa* (presence of abnormal implants), might be related to the observation (Perusquía *et al.*, 1985) that the former preparation stimulated diestrous, estrous, and ovariectomized rat uterine contractility *in vitro*, while the latter had an inhibitory effect.

The crude extract used by Levine *et al.* (1981) [and apparently also that used by Hahn *et al.* (1981a)] was prepared first by extracting *M. tomentosa* leaves with hot water for several hours, and then extracting the filtered tea with a series of organic solvents; the semipurified extract was obtained by silica gel chromatography of the crude extract. The semipurified extract has also been referred to as the "triplet" because gas chromatography had indicated that it contained three components; two of these, zoapatanol and montanol, were subsequently isolated. Guzmán *et al.* (1985) attempted to isolate the third component of the triplet by means of reverse-phase high-performance liquid chromatography. They succeeded in a qualitative separation of the triplet, as well as of a less-purified septuplet, reporting the retention times of zoapatanol, the unknown, and montanol in the septuplet as 4.8, 6.4, and 6.96 min, respectively. Preparative separation of both the triplet and septuplet failed, however; only zoapatanol and montanol peaks were obtained. Since the unknown was shown to be overlapping with montanol, the authors concluded that these two compounds must be structurally related, the unknown easily isomerizing to montanol. They suggested that the double bond in the unknown compound "**9**" might be changing from a tetrasubstituted extension to a position conjugated with the carbonyl.

It now appears, however, that the "triplet" actually contains more than three compounds. From it, in addition to zoapatanol and montanol, Quijano *et al.* (1985a) have isolated tomentol (**10**), zoapatanolide E (**11**), and tomexanthol (**12**). Tomentol, the structure of which was elucidated in this study by means of [^{1}H]NMR, UV, IR, and MS, could be the unknown of Guzmán *et al.* (1985); the isomerization of tomentol to montanol could easily be envisioned, but so might the isomerization of "**9**" to montanol, analogously to the isomerization of zoapatanol to 21-normontanol (**13**) (Marcelle *et al.*, 1985). Quijano *et al.* (1985b), however, noting that zoapatanol and tomentol (not montanol) were the major components of their zoapatle diterpenoid mixture, reinvestigated *M. tomentosa* using mild extraction conditions; their results indicated that montanol was an artefact derived from the isomerization of tomentol. We, too, have observed the isomerization of tomentol to montanol in our laboratory (unpublished observations). Quijano *et al.* (1985a) have also reported that tomentol was as active uterotonically as were the "triplet mixture" and zoapatanol, but gave no further details. To the extent that this might (or might not) correlate with fertility-regulating activity (*vide infra*), variation in biological response to zoapatle preparations could be expected, depending on which isomer(s) predominated, which in turn could depend on which extraction procedures were used. Marcelle *et al.* (1985), for example, found little or no zoapatanol to be extracted with hot water or with hexane; ethanol and ethyl acetate, instead, proved to be useful for this purpose.

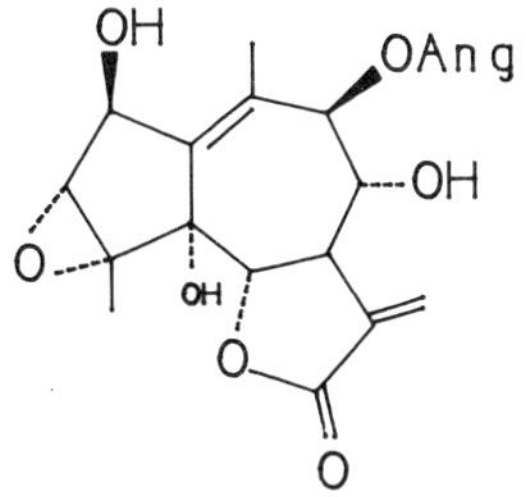

Zoapatanolide E **11**

As already stated above, much of the work concerning zoapatle has involved investigating whether its various components might be able to stimulate uterine contractions *in vitro*. In our laboratory (Fong *et al.*, 1987; Waller *et al.*, 1987), we have been studying the effects of *M. tomentosa* ssp. *tomentosa* and its constituents in the

TABLE I

PREGNANCY-INHIBITING EFFECTS OF *MONTANOA* SPECIES AND CONSTITUENTS THEREOF

Substance tested[a]	Dose[b]	Route of administration	Day(s) *post-coitum*	Species	Results at 3–7 days following treatment	Reference
M. frutescens crude extract	50 mg equivalent of dried leaves	Intrauterine	Day 4	Rat	0/6 rats lacked implants in their extract-treated uterine horns; all had implants in their H_2O-treated horns	Pedrón *et al.* (1985)
M. tomentosa ssp. *tomentosa* crude extract	50 mg equivalent of dried leaves	Intrauterine	Day 4	Rat	6/6 rats had implants in both their extract- and H_2O-treated uterine horns, but the majority of implants in the extract-treated horns reportedly appeared to be abnormal	Pedrón *et al.* (1985)
M. tomentosa						
crude extract	150 mg/kg	p.o.	Days 1–6	Rat	0/5 pregnant	Levine *et al.* (1981)
semipurified extract	300 mg/kg	s.c.	Days 4–6	Hamster	2/8 pregnant	
semipurified extract	100 mg/kg	i.p.	Day 16	Guinea-pig	100% resorbing implants in 5/6 animals	
M. tomentosa						
crude extract	500 mg/kg	i.p.	Day 22	Guinea-pig	8/8 non-viable implants among 3 animals	Hahn *et al.* (1981a)
semipurified extract	228 mg/kg	p.o.	Day 22	Guinea-pig	36/36 non-viable implants among 10 animals	

semipurified extract	100 mg/kg	i.p.	Day 22	Guinea-pig	27/32 non-viable implants among 11 animals	
zoapatanol	150 mg/kg	p.o.	Day 22	Guinea-pig	19/20 non-viable implants among 6 animals	
zoapatanol	100 mg/kg	i.p.	Day 22	Guinea-pig	14/14 non-viable implants among 6 animals	
M. tomentosa ssp. *tomentosa*[c]						D. P. Waller and H. H. S. Fong (unpublished observations, 1985)
EtOAc (E)	500 mg/kg	i.p.	Day 22	Guinea-pig	9/17 implantation sites among 5 animals lacked normal fetuses	
hexane (F)	500 mg/kg	i.p.	Day 22	Guinea-pig	21/27 implantation sites among 6 animals lacked normal fetuses	
EtOAc (A)	500 mg/kg	i.p.	Day 22	Guinea-pig	12/18 implantation sites among 5 animals lacked normal fetuses	
zoapatanolide A	75 mg/kg	i.p.	Day 22	Guinea-pig	11/22 implantation sites among 5 (of 7 treated) surviving animals lacked normal fetuses	
zoapatanol	100 mg/kg	i.p.	Day 43 or 44	Guinea-pig	10/14 implantation sites among 6 animals lacked normal fetuses	

[a] Control data in these studies were approximately 88–100% pregnancy rates/viable implants.
[b] For studies in which more than one dose per route of administration was used, only the maximally effective one is indicated.
[c] See Scheme I for fractionation details.

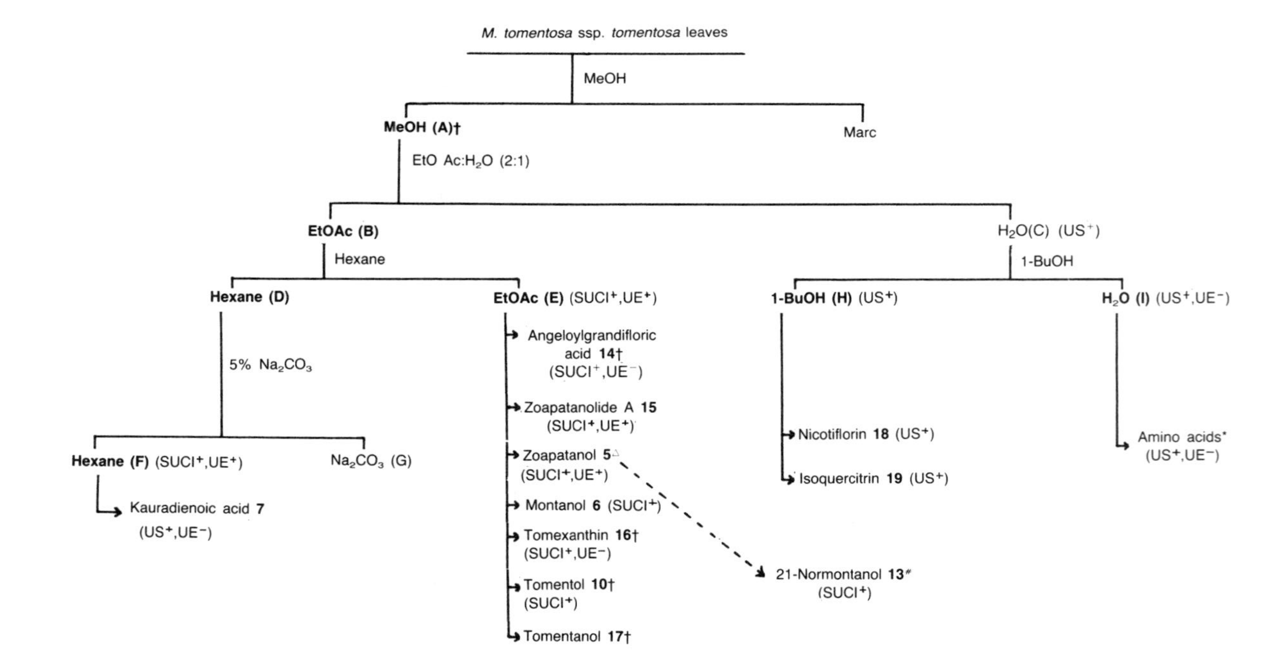

SCHEME I Fractionation scheme and bioassay results for *M. tomentosa* ssp. *tomentosa* leaves and constituents thereof. †For subsequent plant work-up, including the isolation of additional compounds, EtOAc, instead of MeOH, was used for the initial leaf extraction; this extract, shown to be UE^+, is labeled EtOAc (A) in Table I. Compounds indicated † above, were isolated by chromatographic separation of fractions of this extract, after it had been defatted with hexane (Oshima *et al.*, 1986, and unpublished results). *For biological testing, a mixture of amino acids was prepared in the same ratio as had been found in the plant (Waller *et al.*, 1987). △Synthetic (±)-zoapatanol, supplied by K.C. Nicolaou, was used for biological testing. #21-Normontanol was prepared from zoapatanol by heating the latter, or by allowing it to stand in the light at room temperature (Marcelle *et al.*, 1985).

estrogenized guinea-pig uterine strip (Ponce-Monter *et al.*, 1983) and 22-day (or later; see Table I) pregnant guinea-pig (Hahn *et al.*, 1981a) bioassays. Briefly, we have been able to see two different types of change in the spontaneous uterine contraction pattern following *in vitro* application of various zoapatle materials, i.e., either a uterine stimulant (US^+) effect or a spontaneous uterine contraction inhibitory ($SUCI^+$) effect. In the *in vivo* assay, we have watched for the interruption of pregnancy (utero-evacuation) in guinea-pigs, or lack thereof (UE^+ and UE^-, respectively), following single i.p. injections of zoapatle materials.

Our initial zoapatle fractionation scheme, and the various modifications employed, are shown in Scheme I; also indicated are results (+ or −) for the materials that we have tested in the bioassays. As can be seen, the majority of fractions and isolates have been tested *in vitro*; however, some compounds have not yet been tested *in vivo* due to paucity of material. Of the nine materials that have been tested in both systems, four were $SUCI^+$, UE^+, three US^+, UE^-, and two $SUCI^+$,UE^-; further details concerning our positive *in vivo* results are shown in Table I.

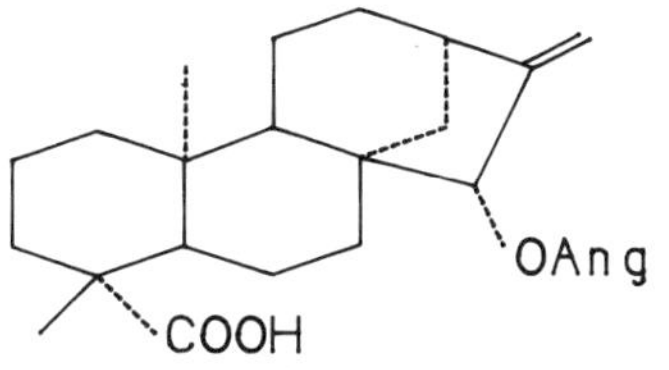

Angeloylgrandifloric acid **14**

At present, we cannot speculate as to why a substance that inhibits spontaneous uterine contractions *in vitro* may be able to interrupt pregnancy when administered *in vivo*, but we do suggest that materials exhibiting such activity *in vitro* might be worth investigating further for possible fertility-regulating activity. Considering the structural

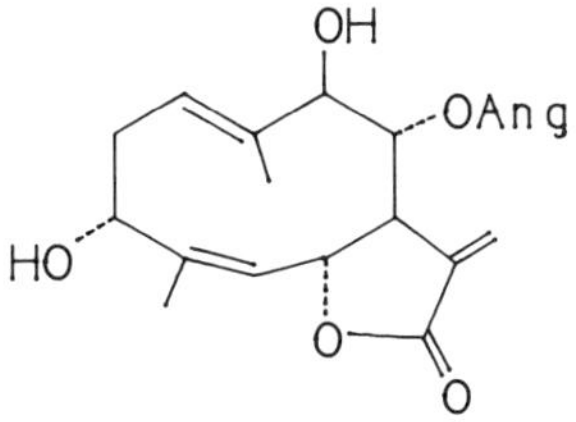

Zoapatanolide A **15**

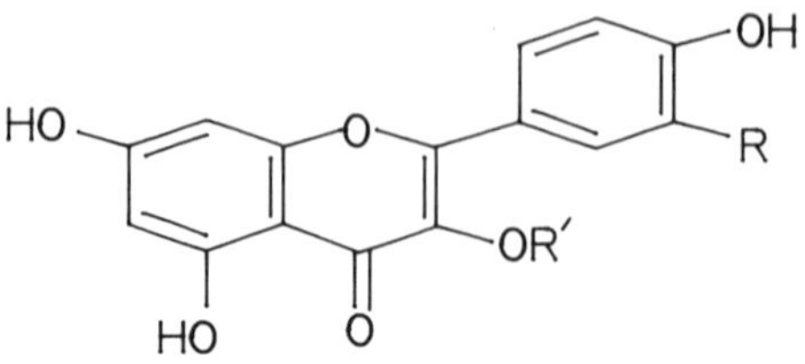

	R	R′
Nicotiflorin **18**	H	Rutinose
Isoquercitrin **19**	OH	Glucose

similarity (just a difference in the position of a double bond) between zoapatanol and 21-normontanol, and among montanol, tomentol, and tomentanol, it will be interesting to see if the SUCI$^+$,UE$^+$ relationship holds true among these five compounds. Hahn *et al.* (1981a), however, have reported montanol to be inactive at 100 mg/kg i.p. in the 22-day guinea-pig assay; only 2/13 implants among six animals were found to be non-viable. Nevertheless, zoapatle extracts apparently do contain at least a couple of constituents that have fertility-regulating activity *in vivo* (see Table I); zoapatle's effects when used ethnomedically may well have been due to the working in concert of the latter compounds, and of others not yet tested pharmacologically (*vide supra*) (see Scheme I).

Of all the analogs of zoapatanol that had been synthesized, 1*RS*,4*SR*,5*RS*-4-(4,8-dimethyl-5-hydroxy-7-nonen-1-yl)-4-methyl-3,8-dioxabicyclo[3.2.1]octane-1-acetic acid (**20**, ORF 13811) was thought to be the most promising (Fong, 1984). Indeed, it has been shown to be effective orally in interrupting pregnancy in various species (Hahn *et al.*, 1984). Specifically, 1/10 and 0/8 mice were

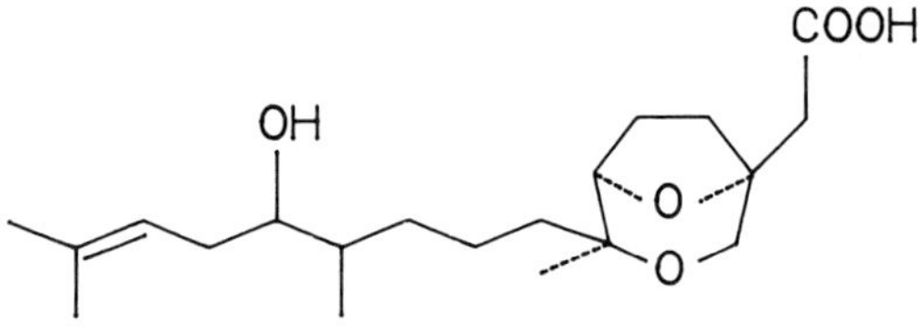

1*RS*, 4*SR*, 5*RS*-4-(4,8-Dimethyl-5-hydroxy-7-nonen-1-yl)-4-methyl-3,8-dioxabicyclo [3.2.1] octane -1-acetic acid **20**

found to be pregnant following administration of 20 mg/kg p.o. on Days 1–6 and 4–6, respectively; the drug was ineffective, however, at 25 mg/kg on Days 1–3. Following single doses (10 mg/kg, rat; 15 mg/kg, mouse) administered on Day 16, 121/121 implants among nine rats and 91/96 implants among eight mice were found to be non-viable. In the guinea-pig, 50 mg/kg p.o. on Day 22 resulted in 12/12 non-viable implants among four animals; comparable results reportedly had been achieved with zoapatanol at a dose of 150 mg/kg p.o. on Day 22 (see Table I). Among three dogs treated with a single dose of 30 mg/kg p.o. between Days 19 and 25, 16/23 implants were found to be non-viable (Hahn *et al.*, 1984). In pregnant baboons, a single dose of 60 mg/kg p.o. administered between Days 30 and 32 resulted in menstruation within 1–3 days and evacuation of uterine contents; 3/4 baboons treated with a single dose of 40 mg/kg between Days 23 and 31 responded similarly within 1–17 days.

Hahn *et al.* (1984) also studied the effects of ORF 13811 on Day-14 (estrous cycle) and Day-22 (pregnancy) guinea-pig uterine strips *in vitro*, and reported the compound to be approximately 30–50 times less potent than $PGF_{2\alpha}$; the authors pointed out, however, that ORF 13811 was effective orally, whereas $PGF_{2\alpha}$ is effective only parenterally. They also found ORF 13811 to be about 15 times less potent than $PGF_{2\alpha}$ in constricting rat aorta *in vitro*; synthetic (±)-zoapatanol similarly had caused constriction of cat coronary arteries (Smith *et al.*, 1981).

Data concerning zoapatle, its constituents, and particularly the synthetic ORF 13811, surely appear promising, and further investigation seems warranted. Even though "non-viable implants", rather than "absence of implants", has been reported when mice, rats, dogs, and guinea-pigs have been treated after the time of implantation, it should be emphasized that effective post-implantation doses (i.e., of ORF 13811) in baboons have resulted in the occurrence of menstruation, evacuation of the uterine contents, and initiation of a new cycle (Hahn *et al.*, 1984).

V. VASICINE

In 1980, Atal (1980) reviewed the work that had been carried out at the Jammu Regional Research Laboratory, on vasicine (**21**), a constituent of *Adhatoda vasica* Nees (Acanthaceae), and also some of the history behind that work. Briefly, an extract of *A. vasica* had been an official preparation included in a number of cough remedies marketed in India. Vasicine, thought by at least some investigators to be the active bronchodilatory agent, was subjected to extensive pharmacological investigation. As a result of the latter, vasicine was reported to have uterine stimulant activity, which activity was lost

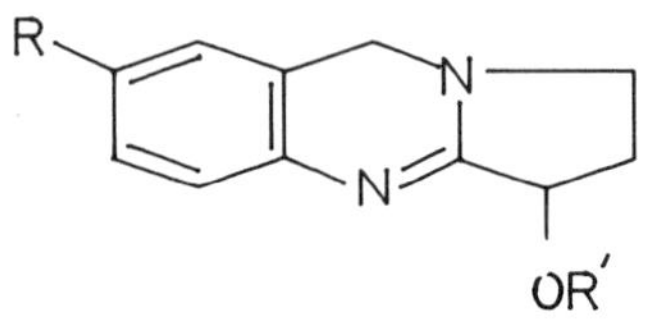

	R	R′
Vasicine **21**	H	H
Vasicine stearate **22**	H	$CH_3(CH_2)_{16}CO$
7-Nitrovasicine **23**	NO_2	H

when vasicine was metabolized by the liver. An examination of the Ayurvedic literature, furthermore, revealed that traditional midwives had used *A. vasica* leaves to control *post-partum* haemorrhage. In 1981, a monograph appeared in *Drugs of the Future* (Arya, 1981), listing vasicine as an oxytocic and abortifacient produced by the Regional Research Lab., Jammu-Tawi, India.

Vasicine, then, has been investigated clinically, as well as experimentally. Much of the most recent work concerning this compound, however, has been directed towards preparing structurally modified compounds that the investigators have hoped might prove to be therapeutically superior to vasicine. The following is a summary of a number of these various studies; further details, and those of others, may be found in the reviews by Atal (1980) and by Jain *et al.* (1984).

Although a large-scale clinical evaluation of vasicine was reported to be in progress (Arya, 1981), the published studies concerning the *in vivo* effects of this compound in humans (as well as in experimental animals) involve relatively few subjects. For example, 9/10 post-maturity or uncoordinated uterine contraction cases were reported to have delivered normally when vasicine (20 mg in 500 ml of dextrose i.v.) was used to induce labor (Atal, 1980); *post-partum* bleeding also reportedly was successfully checked in 60 cases in which vasicine (5–10 mg diluted with saline) was administered i.v. Attempts to induce second-trimester (14–20 weeks) abortion with vasicine have also been reported (Atal, 1980). One of four, 2/3, 5/6, 10/10, 5/5, and 32/34 women aborted following 10, 20, 40, 60, 80, and 100–250 mg of vasicine hydrochloride, respectively, administered intra-amniotically. The administration to abortion interval following doses of 10–80 mg ranged between 72 and 100 hr, and between 30 and 73 hr following doses of 100–250 mg.

Vasicine was only somewhat successful in interrupting pregnancy in laboratory animals, and essentially so only in the later stages (Atal, 1980), related perhaps to its *in vitro* uterine-stimulant (oxytocic) properties. In rats, vasicine was ineffective at doses of 5 or 10 mg/kg i.p. on Days 1–7; half the animals delivered even when given such doses on Days 10–12 or 12–14. Eight of 10 hamsters, however, failed to deliver when given 5 mg/kg of vasicine i.p. on Days 7–9 or 10–12. Group sizes for rabbits under various dosage regimens were extremely small (only 2–4 per group). It would appear, however, that 20 mg/kg of vasicine p.o. on Days 17–19 or 22–24 was not active, while 5–10 mg/kg i.m. on Days 22–24 or 5–10 mg/kg i.p. on Days 17–19 or 22–24 was active; furthermore, 3/3 rabbits aborted when given 2.5 mg/kg i.p. on Days 17–19, whereas 2/4 aborted when administered that dose intramuscularly on those days. Unfortunately, no control data were reported for the experiments involving rabbits.

Vasicine (30 mg/kg i.p.) failed to induce abortion in 6/6 guinea-pigs injected between Days 16 and 20 of pregnancy, and similarly failed in 6/6 others injected between Days 32 and 36 (Gupta *et al.*, 1978); 4/8 guinea-pigs injected with this dose between Days 56 and 60 did abort. Reportedly, however, when a dose of only 10 mg/kg i.p., given at these same stages of pregnancy, was preceded 40 hr earlier by a 50 mcg/kg s.c. dose of estradiol dipropionate, 3/8, 6/12, and 10/12 guinea-pigs aborted, respectively; estradiol dipropionate, alone, did not induce abortion in any of the four guinea-pigs injected at each of these three stages of pregnancy, respectively.

During clinical trials with vasicine, it reportedly was observed that vasicine's activity was less marked following its administration by the i.v. route, and that only partial cervical dilatation could be achieved when it was administered intramuscularly (Jain *et al.*, 1984); even with high doses employed intra-amniotically, the administration to abortion interval could not be reduced below approximately 32 hr. Although further reduction of this interval thus seems doubtful [and this interval length does compare favorably with that for other abortifacients already discussed (*vide supra*)], it would nevertheless be advantageous to develop a vasicine preparation that might be effective by a route other than the intra-amniotic one. Some very preliminary results suggest that esters of vasicine, such as vasicine stearate (**22**), might demonstrate increased bioavailability as compared with the parent compound (Atal, 1980); for a 24-hr period following i.m. injections of vasicine stearate in oil or vasicine hydrochloride in water, blood and amniotic fluid levels remained elevated in rats given the stearate as compared with such levels in those given the hydrochloride.

Other types of structural modification of the vasicine molecule have also been carried out, and the compounds tested *in vitro* for uterine stimulant activity. The activity of 7-nitrovasicine (**23**) was reported to be like that of vasicine (Jain *et al.*, 1983), while that of 1,2,3,9-tetrahydro-6,7-methylenedioxypyrrolo[2,1-*b*]quinazoline (**24**) [a derivative of deoxyvasicine (**25**)] (Chowdhury *et al.*, 1985), and of

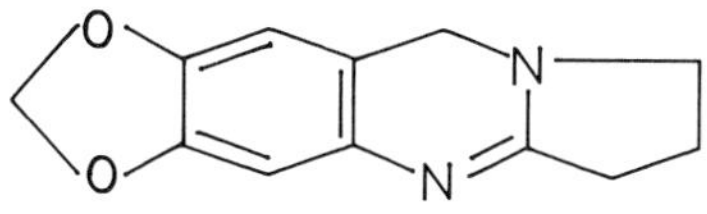

1,2,3,9-Tetrahydro-6,7-methylenedioxy-pyrrolo [2,1-*b*] quinazoline **24**

other deoxyvasicine derivatives (**26–31**) (Jain *et al.*, 1984), was each reportedly greater than that of vasicine. Nevertheless, the last derivative listed (**31**) was reported to be approximately 100-fold *less* potent than was oxytocin, when tested *in vitro* on the estrogenized rat uterus (Rao *et al.*, 1982).

Atal (1980) has argued that one important advantage of vasicine is its inexpensive availability from an abundant natural source, *Adhatoda vasica*. It is unclear, however, whether this compound is indeed currently being used as an abortifacient, and if so, why no further data on this use appear to have been published. Perhaps further toxicological studies are needed? Some toxicity studies reportedly had been carried out in a few laboratory animal species (Atal, 1980), and it was Atal's opinion that the results suggested the drug to be safe. Apparently, however, only one small clinical toxicology study has been published (Wakhloo *et al.*, 1980). Twenty-four volunteers, the majority identified as *post-partum* hospital

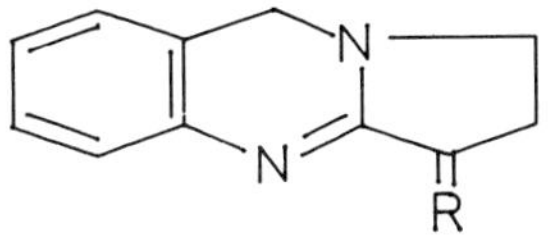

N N R

		R
Deoxyvasicine	**25**	H, H
	26	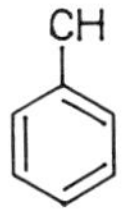 CH
	27	 CH, Cl
	28	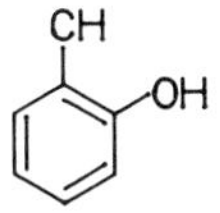 CH, OH
	29	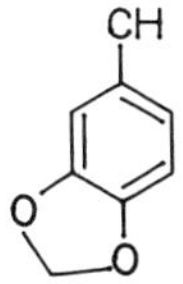 CH, O, O
	30	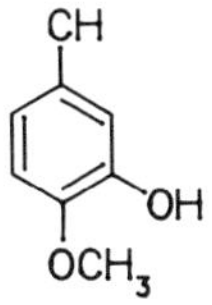 CH, OH, OCH_3
	31	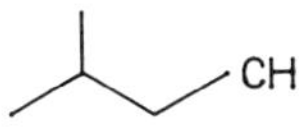 CH

inpatients, received 0.5–16-mg doses of vasicine in 500 ml of saline infused i.v. over 3 hr. Although no undesirable effects were reported, it should be noted that the higest dose tested (16 mg), given to 12 of the volunteers, was clearly below a reliably effective intra-amniotically administered abortifacient dose (*vide supra*).

VI. YUEHCHUKENE

Murraya paniculata L. (Rutaceae), like *Adhatoda vasica*, also has a history of traditional medical use around the time of parturition; specifically, an aqueous decoction of its root reportedly has been used in China to promote labor at term (Kong *et al.*, 1985b). While investigating this plant for possible fertility-regulating effects, Kong *et al.* (1985b) found that a crude chloroform extract, administered p.o. on Days 1–10 *post-coitum*, was able to prevent pregnancy in rats. Further investigation ultimately led to the isolation of a bis-indole alkaloid, yuehchukene (**32**), which was shown to prevent pregnancy in rats when given at a dose of 2.5 mg/kg p.o. on Days 1–2 or 3–4

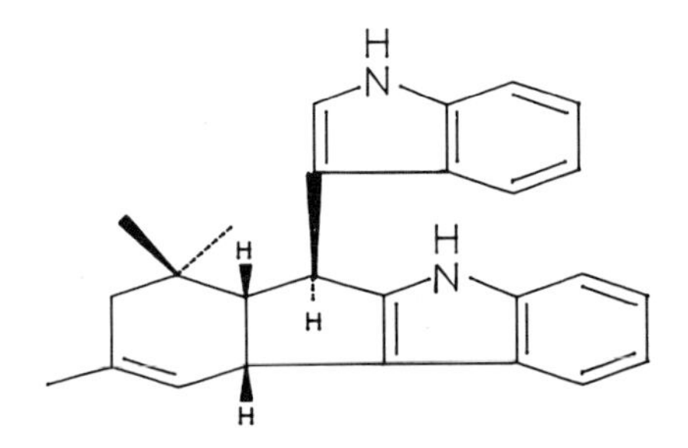

Yuehchukene **32**

(Kong *et al.*, 1985b); similar dosing on Days 5–6 or 7–8 was ineffective, while a single dose of 3 mg/kg p.o. on Day 2, but not on Day 3, effectively prevented pregnancy in 5/5 rats. Yuehchukene reportedly also prevented pregnancy in mice and hamsters (Kong *et al.*, 1985b), but a higher dose was required.

Yuehchukene has also been detected in *M. alata* Drake, *M. exotica* L., and *M. paniculata* var. *omphalocarpa* (Hay.) Tanaka (Kong *et al.*, 1986a). The compound has also been synthesized and the latter shown to possess the same antifertility effect reported for naturally-occurring yuehchukene (Cheng *et al.*, 1985).

Since it is common knowledge that estrogens can prevent implantation (Bingel and Farnsworth, 1980), yuehchukene was subjected to bioassays for estrogenicity (Kong *et al.*, 1985a); it was reported to be inactive in the immature rat uterus and zygote transport bioassays, when tested at fertility-regulating effective doses.

Whether or not yuehchukene might have activity in the human remains to be determined. If so proven, and if it indeed lacks estrogenicity at pregnancy-preventing dosages, its use could be advantageous in women in whom that of estrogens might be contraindicated. Unfortunately, however, if it were shown to act in the human identically to the way in which it acts in the rat, it would have the same disadvantage as do estrogens with respect to the precise timing required, i.e., administration shortly following ovulation, and definitely prior to the time of implantation. Effectiveness, or lack thereof, of agents acting between the occurrences of ovulation and implantation, is relatively easy to determine in laboratory animals such as rats, mice, and hamsters, in which mating and ovulation coincide with hours, but would naturally be difficult to determine in humans.

VII. MISCELLANEOUS POST-COITALLY-ADMINISTERED SUBSTANCES

Even if more continuous dosing (e.g., from Days 10 to 25 of one's menstrual cycle) of an implantation-preventing agent were necessary in the human in order to be sure of covering one's approximated ovulation to (potential) implantation interval, the attractiveness of using an agent that could prevent the occurrence of implantation, rather than one that could only interrupt and undo an already established implantation, should be obvious. Even if such potentially discoverable agents proved to be estrogenic, their ready availability from natural sources could nevertheless give them an advantage over synthetic estrogens in situations in which estrogen use was not contraindicated; whether or not their estrogenic potency would be sufficient to prevent implantation in the human would, of course, be another question needing an answer before their advantages could be considered to be clear.

Numerous plant extracts and constituents thereof, some of which may be estrogenic, have been investigated for post-coital fertility-regulating activity [see, for example, Bingel and Farnsworth (1980), Farnsworth *et al.* (1983), and Kong *et al.* (1986b), and references therein]; dosing regimens, however, have varied from being purely pre-implantational, e.g., Days 1–4 or 1–5 *post-coitum* in mice and rats, respectively, to being both pre- and post-implantational, e.g., Days 1–7 or 1–10 *post-coitum.*

Data concerning the estrogenicity of miroestrol (**33**), a compound isolated from *Pueraria mirifica* Airy-Shaw & Suvatab. (Leguminosae), and preliminary results concerning the estrogenicity and possible antifertility effects of *P. tuberosa* DC., were reviewed previously (Bingel and Farnsworth, 1980), but further studies have since been carried

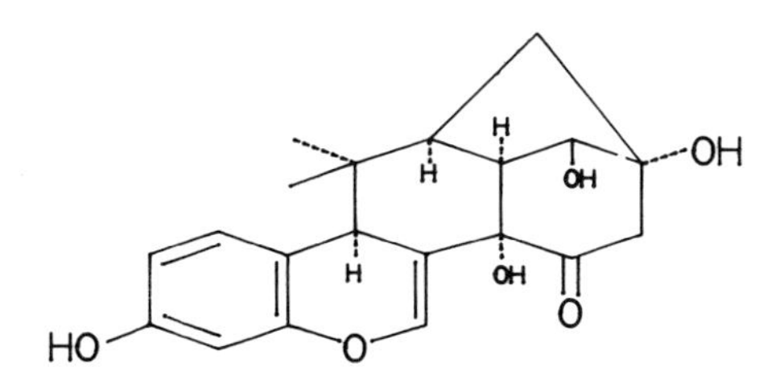

Miroestrol **33**

out on both plants. Feeding powdered *P. tuberosa* tuber (500 mg/day, Days 1–7 *post-coitum*) impaired fertility in 6/6 rats (Mathur *et al.*, 1983); only abnormal/resorptive implantation sites were found, and the vaginal smears contained cornified cells. An alcohol extract of *P. tuberosa* tuber (125 mg/kg p.o., Days 1–7 *post-coitum*) also reportedly exerted an antifertility effect in rats (Chandhoke *et al.*, 1980); non-mated rats, treated with the same or higher doses, reportedly exhibited an anovulatory, persistent estrous condition. Similarly, feeding powdered *P. mirifica* tuber (100 mg/day, Days 1–10 *post-coitum*) prevented implantation in 7/7 rats (Smitasiri *et al.*, 1985); the same dose also was effective if fed only on Days 1–3, but not if fed on Days 4–6 or 7–9. Based on relative uterine weight increases produced in ovariectomized, immature rats, the estrogenic potency of 1 mg of *P. mirifica* dried powder was determined to be equivalent to that of 0.52–0.75 mcg of ethinylestradiol (Smitasiri *et al.*, 1985). *Pueraria mirifica* (5–7% by weight, mixed with commercial feed) also reportedly impaired the development of spermatogenesis when fed to immature coturnix quails (Jesrichai *et al.*, 1985).

Four other plants [*Butea monosperma* (Lamk.) Kuntze (Leguminosae), *Lagenaria breviflora* Robert (Cucurbitaceae), *Lepidium capitatum* Hook. f. & Thoms. (Cruciferae), and *Ferula jaeschkeana* Vatke (Umbelliferae)] have also recently been investigated for post-coital antifertility effects.

Studies reporting post-coital antifertility properties for *B. monosperma*, properties apparently not explainable by estrogenic or antiestrogenic action, were reviewed earlier (Bingel and Farnsworth, 1980), but the extracts that had been investigated were of the flowers. The results of a recent study involving *B. monosperma* seeds, on the other hand, indicated the presence of a post-coitally active constituent, butin (**34**), a flavone that does appear to be weakly estrogenic (Bhargava, 1986). Six of 10 rats treated on Days 1–5

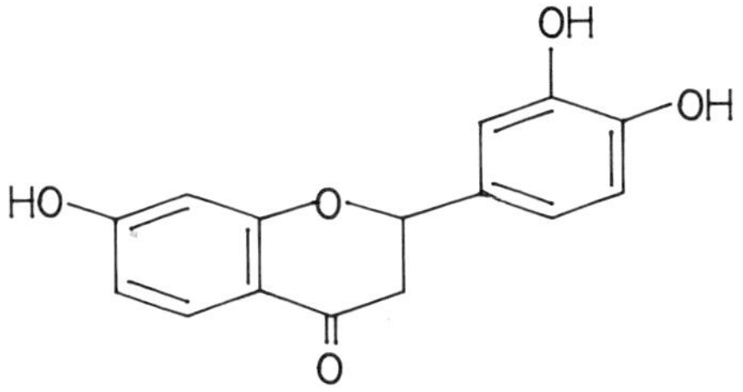

OH
OH
HO
O
O

Butin **34**

post-coitum (20 mg butin/rat/day p.o.) lacked implantation sites on Day 10, and the remaining four failed to deliver at term. Although only 4/10 rats treated with 10 mg/day lacked implantation sites on Day 10, three of the remaining six also failed to deliver at term. One mg of butin/ovariectomized young rat/day × 5 days p.o. was required to double the mean uterine weight over that for the controls, while only 0.1 mcg of estradiol-17β/ovariectomized rat/day × 5 days i.m. was required to do so. A preponderance of cornified cells in the vaginal smears was found following similarly administered doses, respectively, of 10 mg of butin or 1 mcg of estradiol-17β.

A non-toxic dose (20 g-equiv./kg p.o.) of a methanol extract of the dried whole fruit of *L. breviflora* (Elujoba *et al.*, 1985) prevented implantation (4/10 rats pregnant *versus* 10/10 vehicle controls) when administered p.o. on Days 1–10. Methanol extracts of the fresh fruit pulp appeared to be more potent: 2/10 rats pregnant when dosed with only 2.5 g-equiv. (moisture-free)/kg/day, and 0/9 pregnant (plus one death) when dosed with 5 g-equiv./kg/day. The methanol extract of the fresh seed [2 g-equiv. (moisture-free)/kg/day] also was active (0/6 pregnant), but toxic (4 deaths). The compound(s) responsible for this implantation-preventing effect has(ve) not yet been isolated.

An ethanol extract of *L. capitatum* (entire plant) (Singh *et al.*, 1984) was partitioned against hexane, resulting in hexane-soluble and hexane-insoluble fractions. The hexane-soluble fraction (250 mg/kg p.o., Days 1–5) prevented implantation in 18/18 rats, but was inactive at this dose in hamsters. This fraction was also demonstrated to be estrogenic (albeit about 1/10,000 the potency of ethinylestradiol) when tested for uterine weight increase, vaginal opening, and vaginal cornification in ovariectomized, immature rats, and for ability to induce delayed implantation in ovariectomized mice. Isolation of the active compound(s) has not yet been reported.

An ethanol extract of *F. jaeschkeana* (entire plant) (Singh *et al.*, 1985) prevented implantation in rats when administered p.o. at doses of 250 or 500 mg/kg on Days 1–5, as did the serially obtained hexane, benzene, and chloroform fractions at doses of 50 or 100 mg/kg; none of the aforementioned was active in hamsters. Butanol and aqueous fractions, tested only in rats, were inactive at doses of 250 mg/kg. Ferujol (**35**) was isolated from the pooled hexane,

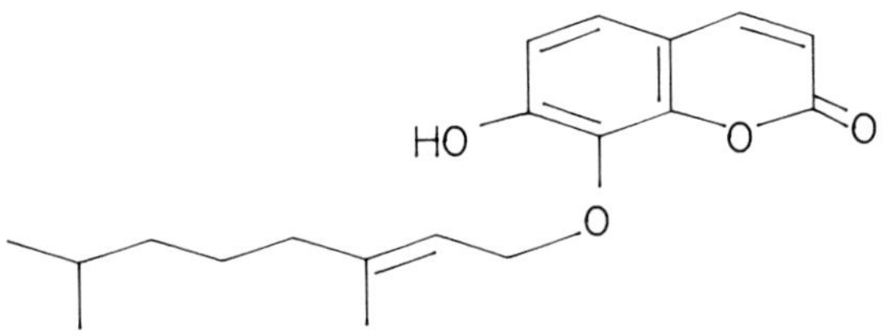

HO O O O

Ferujol **35**

benzene, and chloroform fractions, and subsequently shown to prevent implantation in rats when administered on Day 1 at a dose of 0.6 mg/kg p.o. It reportedly also was active when given on Days 2 or 3 *post-coitum*, but was inactive when given post-coitally to hamsters, even at a dose of 2.4 mg/kg. Ferujol's antifertility effect may be due to its estrogenicity, estimated to be about 1/10 that of ethinylestradiol; the compound tested positive in the same estrogenicity bioassays that Singh *et al.* (1984) had used in testing the hexane-soluble fraction of *L. capitatum*.

In 1980, Bingel and Farnsworth (1980) reviewed available literature concerning post-coital fertility-regulating effects reported for *Embelia ribes* Burm. f. (Myrsinaceae) and its constituent, embelin (**36**), and those reported for *Aristolochia indica* L. (Aristolochiaceae) and its various constituents. Briefly, results obtained with *E. ribes* extracts and/or embelin had varied from negative through equivocal

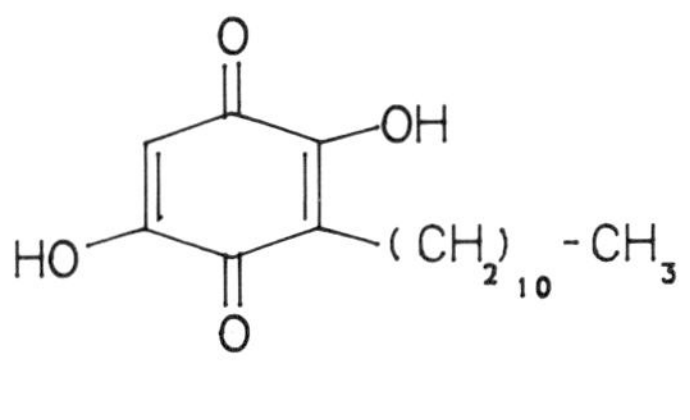

O
OH
HO
$(CH_2)_{10}-CH_3$
O

Embelin **36**

to positive, among different groups of investigators, but not always in a dose-dependent manner. Superficially, some of the *A. indica* constituents had appeared to be active, but clear interpretation of the reported data was not possible when the reader was told, for example, that antifertility effects had been seen in 63.6% of a group containing 10 mice!

The results of more recent studies by three groups of workers would superficially seem to have confirmed the effectiveness of embelin (20–100 mg/kg p.o., in 1% gum tragacanth or in gum acacia, administered on Days 1–5 or 1–7) in preventing implantation in significant numbers of rats (Bhargava *et al.*, 1984; Krishnaswamy and Purushothaman, 1980; Prakash, 1981) and/or in preventing delivery (promoting resorption) in some rats in which implantation had occurred (Bhargava *et al.*, 1984; Krishnaswamy and Purushothaman, 1980). In the study by Krishnaswamy and Purushothaman (1980), however, females reportedly were placed with males during the estrous, rather than proestrous, phase of their cycles. In two of the studies (Bhargava *et al.*, 1984; Krishnaswamy and Purushothaman, 1980) [and possibly in the third (Prakash, 1981) as well, based on the reference cited in its Materials and Methods section], a ratio of 3 females:1 male reportedly was used, but no mention was made as to how the results of such matings were distributed with respect to treatment. In the study by Prakash (1981), apparently significant data, e.g., 3/21 treated rats pregnant *versus* 12/12 controls, reportedly were pooled from three experiments carried out at different time intervals. Provided that the number of animals had been evenly divided among the three experiments, these apparently significant data would probably still remain so; nevertheless, it should be pointed out that the size of the various pooled experimental and control groups in this study ranged between 12 and 21. In our laboratory (unpublished observations), 200 mg/kg of embelin in PVP complex, administered p.o. on Days 1–10, was ineffective [9/9 rats (individually paired with males) pregnant, plus one death]; 300 mg/kg likewise was ineffective (7/7 pregnant, plus one death). Only two of four rats survived a dose of 375 mg/kg; one carried a normal pregnancy while all 15 fetuses of the second rat were degenerating. Some reasons for the discrepant results may involve experimental design. In addition, Bhargava *et al.* (1984), at least, reported using an inbred strain; such could have led to subtle impairment of fertility, although their control delivery rate, 15/16, nevertheless was normal. It may also be pertinent to note here that varying sensitivity of different rat strains to the antifertility effects of gossypol has been reported (Waller *et al.*, 1985).

Further work (Che *et al.*, 1984; Wang and Zheng, 1984) has also been carried out on *Aristolochia* spp. and constituents thereof. Che *et al.* (1984), in our laboratory, showed that an ethanol extract of *A. indica* roots could prevent pregnancy in rats and hamsters when administered p.o. on Days 1–10 and 1–6, respectively; various fractions of this extract, however, were toxic or inactive when tested further in rats. Various constituents of *A. indica* [aristolochic acid-I (**37**), savinin (**38**), aristolic acid (**39**), methyl aristolate (**40**), (12*S*)-7,12-secoishwaran-12-ol (**41**), and *p*-coumaric acid (**42**)] also were tested for post-coital antifertility effects in rats, hamsters, and/or mice, but none was demonstrated to be effective, even when tested in a manner comparable to that reported in the literature to have given positive results. That our total ethanol extract was active, but its isolated components not so, could indicate that a combination of constituents acting together might be necessary for (maximum) activity (Che *et al.*, 1984); this has also been suggested in connection with zoapatle tea and *D. genkwa* elixir (*vide supra*). Likewise, strain differences, and/or usage of doses inadequate for our animal strains, might explain the discrepancy between our negative results and the positive results reported previously in the literature; however, since our animals are obtained from commercial breeding companies, where they presumably are selected and bred for best possible health and reproductive capacity, might it simply not be more difficult to demonstrate effectiveness of a perhaps only weakly active compound in such animals?

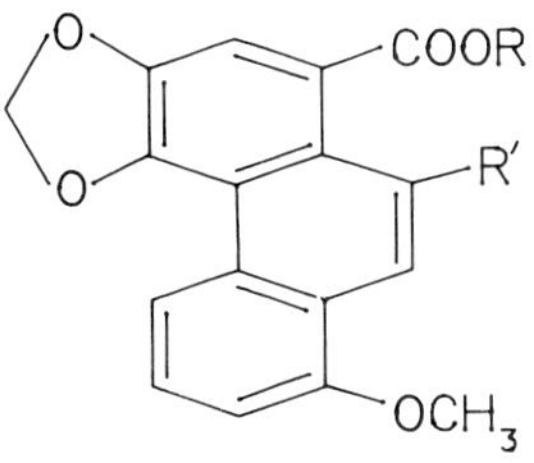

	R	R′
Aristolochic acid-I **37**	H	NO_2
Aristolic acid **39**	H	H
Methyl aristolate **40**	CH_3	H

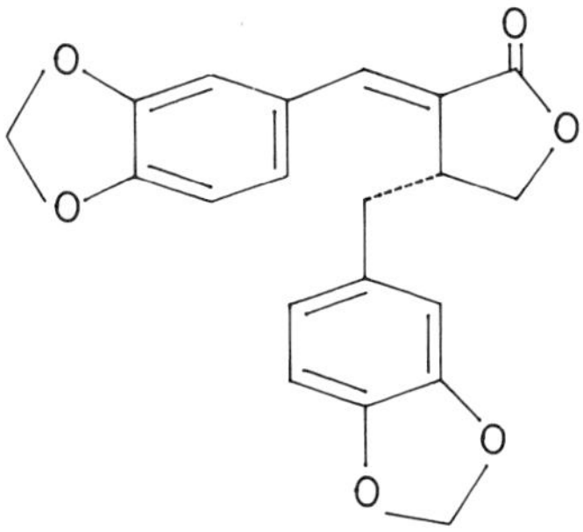

Savinin **38**

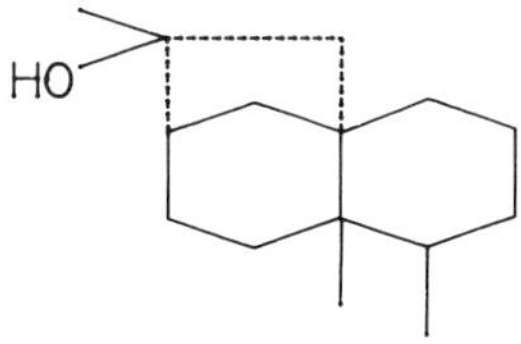

(12*S*)-7,12-Secoishwaran-12-ol **41**

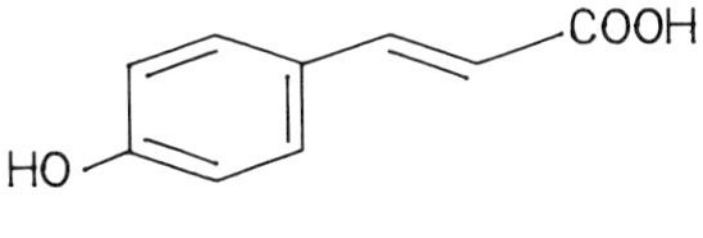

p-Coumaric acid **42**

Aristolochic acid-I [also known as aristolochic acid-A (Mix *et al.*, 1982)] has also been isolated from *A. molissima* Hance. The ethanol extract of this plant, administered p.o. to mice and rats on Days 1–6 (37 g/kg and 54 g/kg, respectively), effectively prevented pregnancy (Wang and Zheng, 1984); aristolochic acid-I also prevented pregnancy in mice (11.1 mg/kg p.o., Days 1–6), but not in rats (20 mg/kg p.o., Days 1–6). It was suggested (Wang and Zheng, 1984) that in the rats, at least, the effectiveness of the extract may have been due to a combination of constituents. In our laboratory (Che *et al.*, 1984), a higher dose of aristolochic acid-I (40 mg/kg p.o., Days 1–10) was lethal to 9/10 rats, with most dying before the dosing regimen could be completed.

Further investigation of aristolochic acid-I showed that it lacked estrogenicity when tested in mice for vaginal cornification and uterine weight increase (Wang and Zheng, 1984). On the other hand, the compound is known to be cytotoxic (Bingel and Farnsworth, 1980), which could explain its effectiveness in causing fetal death when administered intra-amniotically to 14–16-day pregnant rats and 30–45-day pregnant dogs (Wang and Zheng, 1984).

VIII. GOSSYPOL

Waller *et al.* (1985) have recently reviewed the pharmacology of gossypol (**43**), a polyphenolic bis-sesquiterpene, and its status as a male contraceptive. Specifically, they have discussed the history behind its use as such, its various toxicological aspects, and its possible mechanism(s) of antifertility action and/or contributory effects (i.e., its effects on hormones, spermatozoa, sperm enzymes, spermatogenesis, and sperm maturation). They also discussed the

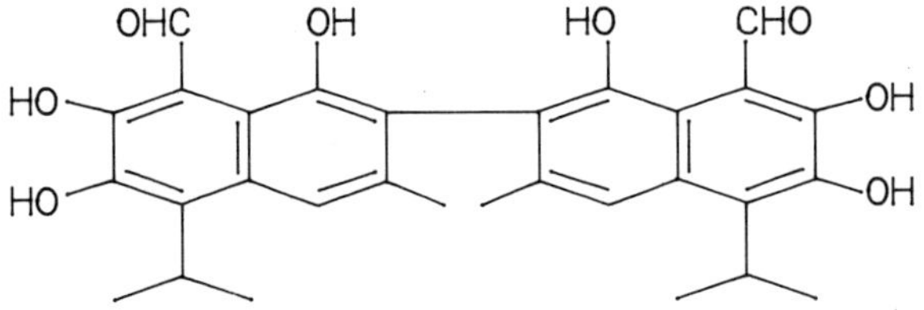

OHC OH HO CHO
HO OH
HO OH

Gossypol **43**

pharmacokinetics of gossypol, and various problems that could be encountered when dosing laboratory animals (and humans) with this compound. Finally, they detailed a number of precautions that should be taken, and several other points that should be considered, by investigators carrying out further, and clearly necessary, antifertility and/or toxicological studies with gossypol. This present review, then, is intended as a brief further update of the pharmacological and toxicological status of gossypol as a male contraceptive; in addition, the results have been summarized of a few recently published studies concerning the effects of gossypol in the female.

It had been well demonstrated that racemic (±)-gossypol, readily available from *Gossypium* species (Malvaceae), could impair fertility (as indicated either by a decreased sperm count or by the results of mating studies) when administered *in vivo* to males of several species [see Waller *et al.* (1985) and references therein]; (+)-gossypol, on the other hand, isolated from *Thespesia populnea* (L.) Soland. ex Correa (Malvaceae), was reported to be lacking in antifertility effects when tested *in vivo* in rats (Wang *et al.*, 1979) and hamsters (Waller *et al.*, 1983a). Such observations naturally led the investigators to speculate that the *in vivo* antifertility effects of (±)-gossypol were probably due to the presence of the levorotary isomer; unfortunately, no plants were known to produce (−)-gossypol.

Recently, however, several methods have been published for resolving the enantiomers of racemic gossypol (Zheng *et al.*, 1985; Matlin *et al.*, 1985; Sampath and Balaram, 1986), and (−)-gossypol has indeed been reported to impair fertility (mating studies) in the small numbers of hamsters (8 mg/kg/day × 40 days p.o.) (Matlin *et al.*, 1985) and rats (15 mg/kg/day × 14 days p.o.) (Wang *et al.*, 1984a) studied. (−)-Gossypol administration (30 mg/kg/day × 7 days) resulted in a significantly lower mean body weight [as compared with that for rats treated with (+)- or (±)-gossypol (30 mg/kg/day × 14 days)] and in the death of 1/5 rats so treated (Wang *et al.*, 1984a); a significantly lower mean body weight had also been noted for hamsters treated with (±)-gossypol (40 mg/kg/day × 54 days) as compared with (+)-gossypol (same dosage regimen), leading Waller *et al.* (1983a) to speculate on the possibly greater toxicity of (−)-gossypol as compared with (+)-gossypol. The latter group, however, also suggested that (−)-gossypol might nevertheless be effective at a lower dose than is racemic gossypol, and that greater toxicity could thus still be avoided. Indeed, even though very few hamsters were used, (−)-gossypol (8 mg/kg/day) appeared to be about as effective as were (±)-gossypol (16 mg/kg/day) and (±)-

gossypol–acetic acid (16 mg/kg/day) (Matlin *et al.*, 1985), and it reportedly did not cause weight loss.

Interestingly, even though (−)-gossypol appears to be the active isomer *in vivo*, such a distinction has not yet been demonstrated with respect to effects of gossypol *in vitro*. It has been shown, for example, that both (+)- and (−)-gossypol are able to immobilize human and hamster sperm completely in the 20-sec and 3-min *in vitro* tests, at the same respective concentrations at which (±)-gossypol immobilizes the sperm of these species (Kim *et al.*, 1984). Minimal *in vitro* concentrations necessary for total immobilization of monkey, rabbit, rat, and mouse sperm, respectively, likewise were identical for (±)- and (+)-gossypol; (−)-gossypol was not tested further due to an insufficiency of material at that time. Likewise, (+)-gossypol, (−)-gossypol, (±)-gossypol, and (±)-gossypol–acetic acid all were shown to inhibit human and hamster testicular LDH-X enzyme activity, when either NADH or α-ketobutyrate was used as substrate (Morris *et al.*, 1986); results were similar for all four substances in each of the four respective experiments, except for a slightly greater potency reportedly demonstrated by (−)-gossypol when tested with hamster LDH-X, with α-ketobutyrate used as substrate. Kim *et al.* (1985), using α-ketoglutarate as substrate, likewise found no difference in the degree of inhibition of rat testicular LDH-X produced by (+)-, (−)-, and (±)-gossypol, but did report a slightly lesser inhibition of hamster testicular LDH-X by (−)-gossypol as compared with that produced by (+)- and (±)-gossypol.

Even the interesting mechanism by which gossypol may impair fertilization at non-spermicidal concentrations (i.e., by inhibiting the conversion of proacrosin to acrosin) was effected *in vitro* by (+)-gossypol as well as by (±)-gossypol (Kennedy *et al.*, 1983). Now that it has become possible to prepare (−)-gossypol, experiments designed to explain the differential effects of (+)- and (−)-gossypol *in vivo* are eagerly and impatiently awaited.

Waller *et al.* (1985) have previously discussed the toxicity of gossypol, and its variability with respect to the species in question. The most serious effect that had been reported in clinical studies was the development of hypokalemia associated with fatigue, lassitude, EEG changes, and occasionally paralysis; it had been suggested, furthermore, that the hypokalemia may have been due to a dietary deficiency of potassium, which was then exacerbated by gossypol. Wang and Yeung (1985) have recently further reviewed the subject of gossypol and hypokalemia; while inadequate dietary intake of potassium does indeed appear to be a predisposing factor, it does not appear to be the sole explanation. The following is a brief summary of their findings.

Wang and Yeung (1985) have concluded that gossypol probably induces renal leakage of potassium and that the most likely mechanism may involve a direct toxic effect of gossypol on the renal tubules. Urinary excretion of potassium was observed to remain high in the presence of a low serum potassium. Nevertheless, there were no other symptoms to suggest mineralocorticoid excess, nor sufficient reports of vomiting or diarrhea to suggest that gastrointestinal loss of potassium could be a major contributory factor. Exactly how gossypol might affect the kidney tubules has not been demonstrated; the results of *in vitro* studies with erythrocytes, however, suggest that potassium loss from the latter cells might be due to an effect of gossypol on the permeability of their cell membranes.

In the various Chinese studies discussed by Wang and Yeung (1985), mean serum potassium levels reportedly tended to fall and level off at the lower limit of normal, during gossypol treatment. The incidence of hypokalemic paralysis, however, seemed to be inversely related to dietary potassium levels as well as to serum potassium levels (Qian, 1981); potassium supplementation apparently could prevent the occurrence of hypokalemic paralysis. Thus, inadequate dietary potassium intake does appear to render a gossypol-treated Chinese patient, at least, susceptible to developing moderate or severe hypokalemia, in turn leading to muscle weakness and paralysis. Wang and Yeung (1985), however, have suggested that genetic susceptibility may also play a role in the development of hypokalemic paralysis, citing both the periodic paralysis of thyrotoxicosis that reportedly occurs only in Orientals, and their own observations of periodic paralysis as a presenting feature in 40% of Chinese patients with primary hyperaldosteronism. Whether or not the potential for developing hypokalemic paralysis, and the ability to prevent it with potassium supplements, are indeed phenomena peculiar only to the Oriental, will not be known for sure until more extensive studies are carried out elsewhere throughout the world.

Waller *et al.* (1985) have indicated that several different versions exist in the literature concerning events leading to the discovery of the antifertility properties of gossypol in the human. According to that related by Zatuchni and Osborn (1981), after exposure for a year or more to crude cottonseed oil prepared by a new method, women had developed amenorrhea, and men had become infertile. After use of the oil had been discontinued, the amenorrheic women resumed having normal menstrual cycles, but many of the exposed men did not immediately regain fertility. The vast majority of studies that have been carried out subsequently have, of course, been concerned with investigating the potential use of gossypol as an antifertility agent for the male. A few recent studies, however, have examined the effects of gossypol in the female. Salient features of some of these are presented below.

As can be seen in Table II, a variety of experimental designs have been used to study the effects of gossypol in the female. Estrous cyclicity seems to be disturbed in rats, but not usually in hamsters. Ovulation also does not appear to be inhibited in hamsters, nor in rats treated on a short-term basis. Pregnancy may be inhibited, but whether or not such inhibition occurs seems to be dosage and/or timing dependent. That gossypol's apparent pregnancy-inhibiting effects may be mediated by a suppression (possibly indirect) of steroid secretion, is suggested by the results of Lin *et al.* (1985) and Wang *et al.* (1984b) (see Table II). Although not all investigators have commented on toxicity, or on lack thereof, side-effects did seem to be at least coincidental, if not causal, to impaired fertility in a few of the studies. Scorza Barcellona *et al.* (1984), furthermore, reportedly found that gossypol acetic acid was toxic to rat fetuses only at dosages that were toxic to the mother rats.

Whether or not gossypol has the potential for usefulness as a systemically administered antifertility agent in the female thus remains unclear at present. Its potential usefulness as a pre-coital, intravaginally applied antifertility agent, on the other hand, continues to show some promise. The results of two studies, one carried out in stumptailed macaques (*Macaca arctoides*) that were douched after every test (Cameron *et al.*, 1982), and the other carried out in women who had previously undergone tubal sterilization (Ratsula *et al.*, 1983), indicated that gossypol, applied intravaginally prior to coitus, could impair the motility of the subsequently ejaculated spermatozoa. In the case of the macaques (Cameron *et al.*, 1982), a glossypol–polyvinylpyrrolidone (PVP) coprecipitate (10–100 mg/ml gelatin) was used; a dose-dependent decrease in sperm motility was observed, approximately 86% being immobilized by the 100-mg/ml preparation. For the clinical experiment (Ratsula *et al.*, 1983), gossypol acetic acid (0.5 mg/ml of gel containing carbopol 940 acrylic acid polymer, triethanol amine, and methylparahydroxybenzoate) was applied intravaginally by means of a 5-ml applicator, approximately 1 hr prior to coitus, and post-coital tests carried out approximately 8 hr *post-coitum*. Compared with their respective control data (post-coital tests carried out following no intravaginal treatment), fewer sperm, all of which lacked motility, were found for 11 of the 15 subjects following gossypol use; for 3/15, both their control and post-treatment tests showed somewhat similar numbers of sperm, but motility of the latter was poor after gossypol use. In contrast, the remaining subject had a better post-coital test with gossypol than she had had without treatment. Results of post-coital tests following use of the vehicle gel alone by five of the subjects, reportedly showed the gel itself to be devoid of spermicidal activity. The investigators had attempted to carry out these post-coital tests approximately 2 weeks prior to expected menstruation (around the time of ovulation) in successive cycles, respectively, for each individual; they were only partially successful in doing so, however (less so in the gossypol treatment cycles that had been carried out first), but this nevertheless did not appear to be able to account for all of the differences observed.

Contraceptive efficacy, of course, could not be determined in either of the last two aforementioned studies. Williams (1980) and Waller *et al.* (1983b), in contrast, have looked at the fertilization rate of ova obtained from small numbers of gonadotropin-treated rabbits, in which gossypol preparations had been applied intravaginally prior to artificial insemination or natural mating, respectively. In the former study (Williams, 1980), gossypol acetic acid was used, mixed with K-Y jelly plus either water or Krebs-Ringer phosphate buffer, and in the latter (Waller *et al.*, 1983b), purified gossypol complexed with PVP (1:4) and mixed with gelatin. Total doses of 1, 2, 5, 10, and 27 mg of gossypol acetic acid reportedly prevented the fertilization of all ova ovulated by each of the five respectively treated rabbits (Williams, 1980). The total doses of gossypol used by Waller *et al.* (1983b), in contrast, ranged downward from 1.25 to 0.2 mg; consequently, zero (or close to zero) percentage fertilization rates were seen only in some instances (D. P. Waller, unpublished observations). That there was even only some inhibition of fertilization is of particular interest, nevertheless, since the gossypol–PVP concentrations selected for intravaginal use were purposely lower than the minimum (4mg/ml) reported to be required for immobilizing 100% of rabbit sperm *in vitro* within 20 sec (Waller *et al.*, 1983b). These observations correlated with others made by this group (Kennedy *et al.*, 1983), i.e., that the percentage of denuded hamster oöcytes penetrated by human sperm *in vitro* could be decreased by prior exposure of the latter to non-spermicidal concentrations of gossypol–PVP. Clearly, larger studies are needed that are designed

TABLE II
EFFECTS REPORTED FOLLOWING GOSSYPOL ACETIC ACID TREATMENT OF THE FEMALE

Species	Dosage regimen[a]	Results	Reference
Hamster	20 mg/kg × 20 days beginning on estrus	2/5 lost considerable weight and stopped having regular 4-day estrous cycles; possible liver toxicity. Normal number of ova reported for remaining animals autopsied on estrus following treatment.	Wu *et al.* (1981)
	10 mg/kg × 20 days beginning on estrus	5/5 continued to have regular 4-day estrous cycles; no significant difference reported between number of ova on estrus following gossypol treatment compared with that following 3% Tween-80 vehicle treatment.	
	10 mg/kg × 40 days beginning on estrus	15/15 continued to have regular 4-day estrous cycles; all 5 that were paired on last day of treatment (proestrus) mated and reportedly had normal pregnancies.	
	5 mg/kg × 76 days beginning on estrus	10/10 continued to have regular 4-day estrous cycles.	
Mouse	75 mg/kg Days 1–15 *post-coitum*	2/8 pregnant; 1 died Day 13; $\bar{x}$ implants = 10.1; 94.5% non-viable; "rough fur" and impaired weight gain reported for gossypol-treated animals.	Sein (1986)
Mouse	80 mg/kg Days 1–13 *post-coitum*	1/10 pregnant; 1 died Day 9; $\bar{x}$ implants = 10.0; 100% non-viable; "unhealthy condition" reported for gossypol-treated animals.	Hahn *et al.* (1981b)

Rat	80 mg/kg on the 3 days prior to expected estrus of 3rd consecutive 4-day cycle	Reportedly did not inhibit ovulation. No data reported concerning side-effects in female rats, but male rats treated with 20 mg/kg/day × 8 weeks reportedly exhibited an impaired weight gain, and 3/10 died.	Hahn *et al.* (1981b)
Rat	30 mg/kg/day, 6 days/week × 8 weeks	Atypical metestrous phase observed in vaginal smears.	Zhou and Lei (1984)
Rat	25 mg/kg × 8 days beginning on proestrus following 3 consecutive 4-day cycles	Cycle irregularity in >75% within 3–5 days; normal estrous cycles resumed 5–7 days after d/c treatment. "Slightly lower" body weight during treatment due to "reduced food consumption"; returned to normal, 4–6 days after d/c treatment.	Lin *et al.* (1985)
	25 mg/kg × 8 days beginning on proestrus following 3 consecutive 4-day cycles	Allowed to mate with fertile males, the same day on which treatment began. Gossypol-treated rats had a reduced number of implantation sites on Day 9 compared with that for 10% EtOH vehicle-treated controls; the former reportedly did not deliver at term.	
	25 mg/kg × 8 days beginning on diestrus following 3 consecutive 4-day cycles	Allowed to mate with fertile males the day after treatment began, i.e., on proestrus. Gossypol-treated rats lacked implantation sites on Day 9. Cause apparently was not an inhibition of ovulation; ovulated ova reportedly were found in similarly treated rats autopsied on Day 2. Other similarly treated rats given, in addition, progesterone 2 mg/rat s.c. on Days 1–22 plus estradiol-17β 0.2 mcg/rat s.c. on Days 5–9, had normal numbers of implantation sites on Day 9 and delivered normal numbers of pups at term. If progesterone treatment was given only through Day 9, normal numbers of implantation sites again were found on Day 9 but significantly fewer pups reportedly were delivered at term.	

Cont'd overleaf

TABLE II Cont'd

EFFECTS REPORTED FOLLOWING GOSSYPOL ACETIC ACID TREATMENT OF THE FEMALE

Species	Dosage regimen[a]	Results	Reference
	25 mg/kg Days 8–15 *post-coitum*	No inhibitory effect on pregnancy reported.	
	25 mg/kg Days 15–22 *post-coitum*	No inhibitory effect on pregnancy reported.	
Rat	25 mg/kg Days 5–15 *post-coitum*	No inhibitory effect on pregnancy reported.	Weinbauer *et al.* (1985)
Rat	80 mg/kg Days 6–9 *post-coitum*	Pregnancy interrupted in 9/10 rats; 8.3% live fetuses. Progesterone (5 mg s.c. Days 6–15) or hCG (25 IU s.c. Days 6–15) reportedly prevented the pregnancy-interrupting effect.	Wang *et al.* (1984b)
	40 mg/kg Days 6–9 *post-coitum*	No inhibitory effect on pregnancy reported.	
Rat	20 mg/kg Days 7–16 *post-coitum*	No inhibitory effect on pregnancy reported.	Beaudoin (1985)

[a]Except for the studies reported by Lin *et al.* (1985), in which the i.m. route was used, gossypol acetic acid was administered p.o. For all studies summarized above, Day 1 of pregnancy (estrus) has been considered the day on which a vaginal plug, or a sperm-positive vaginal smear, was found.

to investigate the contraceptive efficacy of intravaginally applied gossypol; also needed are those designed to investigate whether gossypol might have any toxic effects on the vaginal mucosa, or any undesirable systemic effects if absorbed following intravaginal application.

IX. CONCLUSIONS

Plants do indeed contain a variety of constituents that can be demonstrated to have fertility-regulating properties. How far these substances can be developed and employed for controlling the world's overpopulation, on the one hand, and the individual's unwanted fertility, on the other, remains to be determined.

For the male who wishes to take the responsibility for contraception, gossypol is still the most promising agent from natural sources. However, its *contraceptive* efficacy (as opposed to a mere demonstration of its ability to reduce the sperm count), and its potential for toxicity, must be determined, first in non-human primates and, if results appear promising, then in large-scale clinical studies in non-Orientals, before its ultimate place in the fertility-regulating armamentarium can be predicted.

Brief reports also have appeared in the literature concerning fertility-regulating effects produced in the male by the constituents of plants other than *Gossypium* spp. In comparison with the amount of work that has been carried out on gossypol, however, these reports represent mere pilot studies. A number of these were reviewed in 1982 (Farnsworth and Waller, 1982).

With respect to the female, the most promising results so far appear to be those involving the second-trimester abortifacients, particularly α-trichosanthin and the yuan-hua preparations. α-Trichosanthin, furthermore, reportedly appears to be of some use in the treatment of trophoblastic disorders, i.e., hydatidiform mole (*vide supra*), malignant mole, and choriocarcinoma (Tsao *et al.*, 1986). *In vitro* results of Tsao *et al.* (1986) would seem to support such clinical use. α-Trichosanthin was shown by these workers to have a selective cytotoxic effect on two choriocarcinoma cell lines (and on a mouse melanoma cell line), evident after these cell cultures had been incubated with the drug for 48 hr; two fetal fibroblast cell lines, a squamous carcinoma cell line, and a rat hepatoma cell line, in contrast, were not sensitive to the effects of α-trichosanthin.

Additional/alternative medications for treating trophoblastic disorders are, of course, welcome, and second-trimester abortifacients remain necessary for inducing therapeutic abortions, e.g., because of the mother's health, or because the fetus has been diagnosed as genetically abnormal. For fertility-regulation *per se*, however, an agent that would be effective when taken shortly before the time of one's expected menses and/or when taken shortly after one's first missed menses, would be the ideal; the attractiveness of taking a drug only when needed, as opposed to virtually continuously, which may not always be necessary for a given woman, should be obvious. Clearly, the need for such an agent has already been recognized, and the progesterone antagonist, RU 486 [17β-hydroxy-11β-(4-dimethylaminophenyl)-17α-(prop-1-ynyl)-estra-4,9-dien-3-one], may help to fill that need [see Couzinet *et al.* (1986) and Nieman *et al.* (1987), and references therein]. This synthetic drug interrupted pregnancy in 85/100 women given total doses of 400–800 mg p.o. over 2–4 days, beginning within 10 days of their first missed menstrual period (Couzinet *et al.*, 1986). This drug (10 mg/kg p.o.) also induced menses within 72 hr when given on Day 7 of the luteal phase to six women who were not at risk of pregnancy (Nieman *et al.*, 1987); bleeding likewise was induced within 72 hr in six other similarly treated women, even though the latter were also given hCG (2000 IU/day i.m.) on Days 6–8 of the luteal phase. RU 486 (5 mg/kg i.m.) also reportedly prevented pregnancy when administered to rhesus monkeys on Day 25 of a fecund cycle, i.e., a cycle of 25–33 days in which mating had occurred (Nieman *et al.*, 1987).

The development of RU 486, however, should not be looked at as a final answer, but rather should serve to stimulate investigation in greater depth, of the various plant extracts and constituents discussed above that appear to have exhibited post-coital/pregnancy interrupting effects, as well as continued exploration of the plant kingdom for additional substances that may act similarly. It may well be, for example, that a standardized zoapatle tea containing an appropriate mixture of sufficient quantities of UE^{+} and/or $SUCI^{+}$ constituents, as well as US^{+} constituents, could be developed as a clinically useful, early pregnancy interrupting preparation. Likewise, even though at least some of the plant materials that have been reported to show post-coital, pregnancy-preventing activity in small laboratory animals, may be acting by virtue of being estrogenic, this, if true, should not impede their further study. If proven active in the human, an estrogen from a local plant source could prove invaluable as a fertility-regulating agent in a population that otherwise might not have access to synthetic estrogens.

A contraceptive, then, even the theoretically most effective one, will be effective only if used, and only if used correctly; furthermore, it will be used only if acceptable to and accessible by the user. For at least some of the world's population, "acceptable" and "accessible" may correlate best with the use of medicinal plants. Consequently, leads such as the ones discussed in this chapter should continue to be investigated, the ultimate goal not being an unattainable one, i.e., one (or a few) effective fertility-regulating agent(s) for all, but rather a potentially attainable one, i.e., a diverse armamentarium of effective fertility-regulating agents among which everyone can find at least one that is both acceptable to, and accessible by, himself/herself.

ACKNOWLEDGEMENTS

The authors wish to thank Dr. C. T. Che, Ms. X. Dong, and Dr. S. M. Wong for translating the articles that had been published in Chinese, and Drs. G. Blasko, G. A. Cordell, and D. P. Waller for helpful suggestions and advice.

The studies that have been carried out in our laboratory on *M. tomentosa* ssp. *tomentosa*, *A. indica*, embelin, and gossypol were supported in part by the Special Programme of Research, Development, and Research Training of the World Health Organization (HRP Projects 77918c and 82027).

REFERENCES

Arya, V. P. (1981). *Drugs of the Future* **6**, 373.

Atal, C. K. (1980). "Chemistry and Pharmacology of Vasicine–a New Oxytocic and Abortifacient", 155 pp. Regional Research Laboratory, Jammu-Tawi.

Beaudoin, A. R. (1985). *Teratology* **32**, 251.

Bhargava, S. K. (1986). *J. Ethnopharmacol.* **18**, 95.

Bhargava, S. K., Dixit, V. P., and Khanna, P. (1984). *Fitoterapia* **55**, 302.

Bingel, A. S., and Farnsworth, N. R. (1980). *In* "Progress in Hormone Biochemistry and Pharmacology, Vol. 1" (M. Briggs and A. Corbin, eds.), pp. 149–225. Eden Press, Westmount.

Cameron, S. M., Waller, D. P., and Zaneveld, L. J. D. (1982). *Fert. Steril.* **37**, 273.

Chan, W. Y., Tam, P. P. L., Choi, H. L., Ng, T. B., and Yeung, H. W. (1986). *Contraception* **34**, 537.

Chan, W. Y., Tam, P. P. L., and Yeung, H. W. (1984). *Contraception* **29**, 91.

Chandhoke, N., Gupta, S., Daftari, P., Dhar, S. K., and Atal, C. K. (1980). *Indian J. Pharmacol.* **12**, 57.

Chang, M. C., Saksena, S. K., Lau, I. F., and Wang, Y. H. (1979). *Contraception* **19**, 175.

Che, C. T., Ahmed, M. S., Kang, S. S., Waller, D. P., Bingel, A. S., Martin, A., Rajamahendran, P., Bunyapraphatsara, N., Lankin, D. C., Cordell, G. A., Soejarto, D. D., Wijesekera, R. O. B., and Fong, H. H. S. (1984). *J. Nat. Prod.* **47**, 331.

Chen, H. L., Song, J. F., and Tao, Z. J. (1984). *Acta Physiol. Sin.* **36**, 388.

Cheng, K. F., Kong, Y. C., and Chan, T. Y. (1985). *J. Chem. Soc., Chem. Commun.* **1985**, 48.

Chowdhury, B. K., Afolabi, E. O., Osuide, G., and Sokomba, E. N. (1985). *Indian J. Chem.* **24B**, 789.

Couzinet, B., Le Strat, N., Ulmann, A., Baulieu, E. E., and Schaison G. (1986). *N. Engl. J. Med.* **315**, 1565.

Elujoba, A. A., Olagbende, S. O., and Adesina, S. K. (1985). *J. Ethnopharmacol.* **13**, 281.

Estrada, A. V., Enríquez, R. G., Lozoya, X., Bejar, E., Girón, H., Ponce-Monter, H., and Gallegos, A. J. (1983). *Contraception* **27**, 227.

Farnsworth, N. R., Fong, H. H. S., and Diczfalusy, E. (1983). *In* "Research on the Regulation of Human Fertility, Vol. 2" (E. Diczfalusy and A. Diczfalusy, eds.), pp. 776–809. Scriptor, Copenhagen.

Farnsworth, N. R., and Waller, D. P. (1982). *Res. Front. Fert. Reg.* **2**(1), 1.

Fong, H. H. S. (1984). *In* "Natural Products and Drug Development, Alfred Benzon Symposium 20" (P. Krogsgaard-Larsen, S. Brøgger Christensen, and H. Kofod, eds.), pp. 355–368, Munksgaard, Copenhagen.

Fong, H. H. S., Oshima, Y., Martin, A., Kasnick, S., Zhu, Y., Konno, C., Tu, Z., Lin, Z., Soejarto, D. D., Cordell, G. A., and Waller, D. P. (1987). *In* "Proc. Int. Symp. Trad. Med. Mod. Pharmacol."

Gallegos, A. J. (1983). *Contraception* **27**, 211.

Gallegos, A. J. (1985). *Contraception* **31**, 487.

Gupta, O. P., Anand, K. K., Ghatak, B. J. R., and Atal, C. K. (1978). *Indian J. Exp. Biol.* **16**, 1075.

Guzmán, A., Gallegos, A. J., García-de La Mora, G., and Flores-Moreno, J. M. (1985). *Arch. Invest. Méd. (Méx.)* **16**, 209.

Hahn, D. W., Ericson, E. W., Lai, M. T., and Probst, A. (1981a). *Contraception* **23**, 133.

Hahn, D. W., Rusticus, C., Probst, A., Homm, R., and Johnson, A. N. (1981b). *Contraception* **24**, 97.

Hahn, D. W., Tobia, A. J., Rosenthale, M. E., and McGuire, J. L. (1984). *Contraception* **30**, 39.

Huang, B. (1982). *Tianjin Med. J.* **10**, 284.

Jain, M. P., Atal, C. K., Bandhyopadhyay, R., and Nagavi, B. G. (1983). *Indian J. Pharm. Sci.* **45**, 178.

Jain, M. P., Gupta, V. N., and Atal, C. K. (1984). *Indian Drugs* **21**, 313.

Jesrichai, S., Anuntalabhochai, S., Sinchaisri, T., and Smitasiri, Y. (1985). "Abstr., 11th Conf. Sci. Tech., Kasetsart Univ., Bangkok, Oct. 24–26," pp. 238–239.

Jin, Y. C. (1985). *In* "Advances in Chinese Medicinal Materials Research" (H. M. Chang, H. W. Yeung, W. W. Tso, and A. Koo, eds.), pp. 319–326. World Scientific Publ. Co., Singapore.

Jin, Y. C., Ho, C. C., Wu, I. E., Chen, Z. R., Tong, S. M., Zhou, Y. F., and Wang, D. Z. (1981). *Reprod. Contracep.* **1**, 19.

Kennedy, W. P., Van Der Ven, H. H., Straus, J. W., Bhattacharyya, A. K., Waller, D. P., Zaneveld, L. J. D., and Polakoski, K. L. (1983). *Biol. Reprod.* **29**, 999.

Kim, I. C., Waller, D. P., and Fong, H. H. S. (1985). *J. Androl.* **6**, 344.

Kim, I. C., Waller, D. P., Marcelle, G. B., Cordell, G. A., Fong, H. H. S., Pirkle, W. H., Pilla, L., and Matlin, S. A. (1984). *Contraception* **30**, 253.

Kong, Y. C., Cheng, K. F., Cambie, R. C., and Waterman, P. G. (1985a). *J. Chem. Soc., Chem. Commun.* **1985**, 47.

Kong, Y. C., Ng, K. H., But, P. P. H., Li, Q., Yu, S. X., Zhang, H. T., Cheng, K. F.,

Soejarto, D. D., Kan, W. S., and Waterman, P. G. (1986a). *J. Ethnopharmacol.* **15**, 195.

Kong, Y. C., Ng, K. H., Wat, K. H., Wong, A., Saxena, I. F., Cheng, K. F., But, P. P. H., and Chang, H. T. (1985b). *Planta Med.* **1985**, 304.

Kong, Y. C., Xie, J. X., and But, P. P. H. (1986b). *J. Ethnopharmacol.* **15**, 1.

Krishnaswamy, M., and Purushothaman, K. K. (1980). *Indian J. Exp. Biol.* **18**, 1359.

Landgren, B. M., Aedo, A. R., Hagenfeldt, K., and Diczfalusy, E. (1979). *Am. J. Obstet. Gynecol.* **135**, 480.

Lau, I. F., Saksena, S. K., and Chang, M. C. (1980). *Contraception* **21**, 77.

Law, L. K., Tam, P. P. L., and Yeung, H. W. (1983). *J. Reprod. Fert.* **69**, 597.

H. P. Lei, Z. Y. Song, J. T. Zhang, G. Z. Liu, and H. Z. Tian, eds.) pp. 100–105, Chinese Pharmacology Society and Chinese Association for Science and Technology, Beijing.

Levine, S. D., Hahn, D. W., Cotter, M. L., Greenslade, F. C., Kanojia, R. M., Pasquale, S. A., Wachter, M., and McGuire, J. L. (1981). *J. Reprod. Med.* **26**, 524.

Lin, Y. C., Fukaya, T., Rikihisa, Y., and Walton, A. (1985). *Life Sci.* **37**, 39.

Lin, Z. M., Zhu, M. K., Pang, D. W., Jiang, X. J., Liu, M. Z., Ding, G. S., and Yang, B. Y. (1981). *In* "Recent Advances in Fertility Regulation" (C. F. Chang, D. Griffin, and A. Woolman, eds.), pp. 319–329. Atar S. A., Geneva.

Liu, G., Liu, F., Li, Y., and Yu, S. (1985). *In* "Advances in Chinese Medicinal Materials Research" (H. M. Chang, H. W. Yeung, W. W. Tso, and A. Koo, eds.), pp. 327–333. World Scientific Publ. Co., Singapore.

Lozoya, X., Enríquez, R. G., Bejar, E., Estrada, A. V., Girón, H., Ponce-Monter, H., and Gallegos, A. J. (1983). *Contraception* **27**, 267.

Marcelle, G. B., Bunyapraphatsara, N., Cordell, G. A., Fong, H. H. S., Nicolaou, K. C., and Zipkin, R. E. (1985). *J. Nat. Prod.* **48**, 739.

Mathur, R., Sazena, V., and Prakash, A. O. (1983). *IRCS Med. Sci.* **11**, 522.

Matlin, S. A., Zhou, R., Bialy, G., Blye, R. P., Naqvi, R. H., and Lindberg, M. C. (1985). *Contraception* **31**, 141.

Mix, D. B., Guinaudeau, H., and Shamma, M. (1982). *J. Nat. Prod.* **45**, 657.

Morris, I. D., Higgins, C., and Matlin, S. A. (1986). *J. Reprod. Fert.* **77**, 607.

Nieman, L. K., Choate, T. M., Chrousos, G. P., Healy, D. L., Morin, M., Renquist, D., Merriam, G. R., Spitz, I. M., Bardin, C. W., Baulieu, E. E., and Loriaux, D. L. (1987). *N. Engl. J. Med.* **316**, 187.

Oshima, Y., Cordell, G. A., and Fong, H. H. S. (1986). *Phytochemistry* **25**, 2567.

Pedrón, N., Estrada, A. V., Ponce-Monter, H., Valencia, A., Guzmán, A., and Gallegos, A. J. (1985). *Contraception* **31**, 499.

Perusquía, M., Sánchez, E., Ponce-Monter, H., Estrada, A. V., Pedrón, N., Valencia, A., Guzmán, A., and Gallegos, A. J. (1985). *Contraception* **31**, 543.

Ponce-Monter, H., Girón, H., Lozoya, X., Enríquez, R. G., Bejar, E., Estrada, A. V., and Gallegos, A. J. (1983). *Contraception* **27**, 239.

Prakash, A. O. (1981). *Planta Med.* **41**, 259.

Qian, S. Z. (1981). *In* "Recent Advances in Fertility Regulation" (C. F. Chang, D. Griffin, and A. Woolman, eds.), pp. 152–159. Atar, S.A., Geneva.

Quijano, L., Calderón, J. S., Gómez-Garibay, F., Rosario, V., and Ríos, T. (1985a). *Phytochemistry* **24**, 2337.

Quijano, L., Calderón, J. S., Gómez-Garibay, F., Rosario, V., and Riós, T. (1985b). *Phytochemistry* **24**, 2741.

Rao, M. N. A., Krishnan, S., Jain, M. P., and Anand, K. K. (1982). *Indian J. Pharm. Sci.* **44**, 151.

Ratsula, K., Haukkamaa, M., Wichmann, K., and Luukkainen, T. (1983). *Contraception* **27**, 571.

Sampath, D. S., and Balaram, P. (1986). *J. Chem. Soc., Chem. Commun.* **1986**, 649.

Scorza Barcellona, P., Campana, A., and De Martino, C. (1984). *IRCS Med. Sci.* **12**, 19;

Chem. Abstr. (1984) **100**, 168428c.

Sein, G. M. (1986). *Am. J. Chin. Med.* **14**, 110.

Sentíes, L., and Amayo, R. (1964). *Gac. Méd. Méx.* **94**, 343.

Singh, M. M., Gupta, D. N., Wadhwa, V., Jain, G. K., Khanna, N. M., and Kamboj, V. P. (1985). *Planta Med.* **1985**, 268.

Singh, M. M., Wadhwa, V., Gupta, D. N., Pal, R., Khanna, N. M., and Kamboj, V. P. (1984). *Planta Med.* **1984**, 154.

Smitasiri, Y., Junyatum, U., Songjitsawad, A., Sripromma, P., Trisrisilp, S., and Anuntalabhochai, S. (1985). Abstr. 11th Conf. Sci. Tech., Kasetsart Univ., Bangkok, Oct. 24–26, pp. 338–339.

Smith, J. B., Smith III, E. F., Lefer, A. M., and Nicolaou, K. C. (1981). *Life Sci.* **28**, 2743.

Tam, P. P. L., Chan, W. Y., and Yeung, H. W. (1984a). *J. Reprod. Fert.* **71**, 567.

Tam, P. P. L., Chan, W. Y., and Yeung, H. W. (1985). *In* "Advances in Chinese Medicinal Materials Research" (H. M. Chang, H. W. Yeung, W. W. Tso, and A. Koo, eds.), pp. 335–350. World Scientific Publ. Co., Singapore.

Tam, P. P. L., Law, L. K., and Yeung, H. W. (1984b). *J. Reprod. Fert.* **71**, 33.

Tsao, S. W., Yan, K. T., and Yeung, H. W. (1986). *Toxicon* **24**, 831.

Wakhloo, R. L., Kaul, G., Gupta, O. P., and Atal, C. K. (1980). *Indian J. Pharmacol.* **12**, 129.

Waller, D. P., Bunyapraphatsara, N., Martin, A., Vournazos, C. J., Ahmed, M. S., Soejarto, D. D., Cordell, G. A., Fong, H. H. S., Russell, L. D., and Malone, J. P. (1983a). *J. Androl.* **4**, 276.

Waller, D. P., Martin, A., Oshima, Y., and Fong, H. H. S. (1987). *Contraception* **35**, 147.

Waller, D. P., Martin, A., and Vournazos, C. (1983b). *J. Androl.* **4**, 37.

Waller, D. P., Zaneveld, L. J. D., and Farnsworth, N. R. (1985). *In* "Economic and Medicinal Plant Research, Vol. 1" (H. Wagner, H. Hikino, and N. R. Farnsworth, eds.), pp. 87–112. Academic Press, London.

Wang, C., and Yeung, R. T. T. (1985). *Contraception* **32**, 237.

Wang, C. R., Huang, H. Z., Xu, R. S., Dou, Y. Y., Wu, X. C., Li, Y., and Ouyang, S. H. (1982). *Acta Chim. Sin.* **40**, 835.

Wang, M. Z., Liu, J. S., Song, L. L., Xiang, B. R., and An, D. K. (1986). *Acta Pharm. Sin.* **21**, 119.

Wang, N. G., Guan, M. Z., and Lei, H. P. (1984a). *Acta Pharm. Sin.* **19**, 932.

Wang, N. G., Li, F. Y., Li, H. P., and Lei, H. P. (1984b). *Acta Pharm. Sin.* **19**, 808.

Wang, Q. W., Fu, Y. C., and Zhong, H. (1983). *Chin. J. Obstet. Gynecol.* **18**, 154.

Wang, W. H., and Zheng, J. H. (1984). *Acta Pharm. Sin.* **19**, 405.

Wang, Y., Luo, Y., and Tang, X. (1979). *Acta Pharm. Sin.* **14**, 662.

Weinbauer, G. F., Kalla, N. R., and Frick, J. (1985). *In* "Gossypol" (S. J. Segal, ed.), pp. 79–88. Plenum Publ. Corp., New York.

Williams, W. L. (1980). *Contraception* **22**, 659.

Wu, Y. M., Chappel, S. C., and Flickinger, G. L. (1981). *Contraception* **24**, 259.

Yeung, H. W., Li, W. W., Law, L. K., and Chan, W. Y. (1985). *In* "Advances in Chinese Medicinal Materials Research" (H. M. Chang, H. W. Yeung, W. W. Tso, and A. Koo, eds.), pp. 311–318. World Scientific Publ. Co., Singapore.

Zatuchni, G. I., and Osborn, C. K. (1981). *Res. Front. Fert. Reg.* **1**(4), 1.

Zheng, D. K., Si, Y. K., Meng, J. K., Zhou, J., and Huang, L. (1985). *J. Chem. Soc., Chem. Commun.* **1985**, 168.

Zhong, X. X., and Wang, D. T. (1983). *Reprod. Contracep.* **3**, 24.

Zhou, L. F., and Lei, H. P. (1984). *Acta Pharm. Sin.* **19**, 220.

Zhou, M. H., Li, Q., Shu, H. D., Bao, Y. M., and Chu, Y. H. (1982). *Acta Pharm. Sin.* **17**, 176.

5

Recent Developments in the Chemistry of Plant-derived Anticancer Agents

GÁBOR BLASKÓ*
GEOFFREY A. CORDELL
Program for Collaborative Research in the Pharmaceutical Sciences, College of Pharmacy, University of Illinois at Chicago. Chicago, Illinois 60612, U.S.A.

I. INTRODUCTION

Natural products exhibiting antitumor activity continue to be the subject of extensive research aimed at the development of drugs for the treatment of different human tumors. Several comprehensive

*Permanent address, Central Research Institute for Chemistry, Hungarian Academy of Sciences, Budapest, P.O. Box 17, H-1525, Hungary

ECONOMIC AND MEDICINAL PLANT RESEARCH VOLUME 2
ISBN 0-12-730063-5

reviews (Cassady and Douros, 1980; Cassady *et al.*, 1981; Cordell, 1977; Cordell and Saxton, 1981; Pettit, 1977; Pettit and Cragg, 1978; Pettit and Ode, 1979; Suffness and Cordell, 1985; Taylor and Farnsworth, 1975) summarizing the advances in the chemistry and biology of plant-derived anticancer agents are available.

In this chapter we would like to review the literature published since 1984 on antitumor alkaloids and on the development of dimeric indole alkaloids and non-alkaloidal plant ingredients in cancer therapy since 1980 as a continuation of previous reviews respectively. Studies on sesquiterpene lactones and diterpenes have been omitted from discussion since they would appear to have limited potential for future development. In general, the compounds under discussion can be divided into two major groups; the alkaloids and the non-alkaloids.

A. ANTITUMOR ALKALOIDS

a. Taxus Alkaloids

The alkaloid taxol (**1**) isolated from several *Taxus* species including *T. brevifolia* Nutt., *T. baccata* Barron var. *barroni*, and *T. cuspidata* Sieb. & Zucc. displays very good activity against the B16 melanoma and MX-1 mammary xenograft systems, and shows moderate activity against the L1210, P388, and P1534 leukemia systems, and the CX-1 colon, LX-1 lung xenografts. Taxol (**1**) is effective by a mechanism different from that of any other known anticancer drug that usually acts on DNA, RNA, or protein synthesis. Taxol (**1**) proved to be a mitotic inhibitor that promotes the assembly of calf-brain microtubules, which then become resistant to depolymerization.

Approximately 30 naturally occurring taxane derivatives are presently known (Miller, 1980), among which the bicyclo[6,5,0]dode-

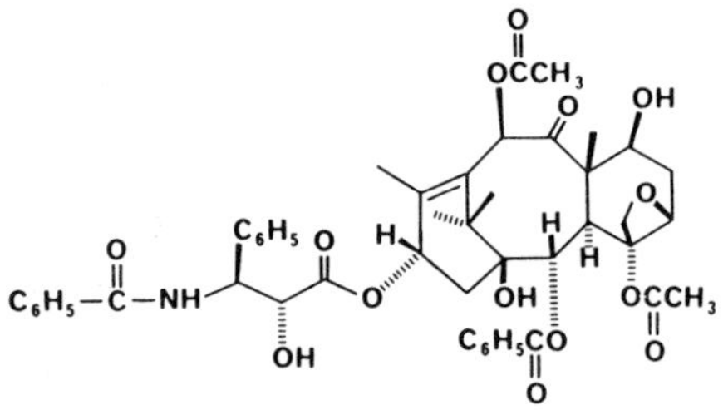

$C_6H_5-C(=O)-NH$ … C_6H_5 … O … OH … $OCCH_3$ … C_6H_5CO

1 Taxol

cane diterpene-type skeleton is common to all except two. Due to the low alkaloid content of the different *Taxus* species and the increasing need for taxane derivatives for advanced pharmacologic investigations, the total synthesis of this type of natural product has attracted attention. However, even the formation of the tricyclic taxane skeleton has proved to be extremely difficult and the total synthesis of taxol still represents a substantial and challenging problem. The first successful attempt (Sakan and Craven, 1983) to obtain a taxane model system utilized an intramolecular Diels–Alder cyclization process as the key reaction step. To facilitate the desired ring closure of substrate **8**, obtained by a linear synthetic pathway as shown in Scheme 2, the five carbon atoms of ring A were rigidly positioned in a boat conformation. The cyclization of **8** in the presence of dimethylaluminum chloride resulted in a 95:5 mixture of products **9** and **10**, respectively. The predominant product **9** was the C-8 epimer of the expected **10** taxane model. Interestingly, a remarkable difference in the stereochemical outcome was observed between the catalysed and uncatalysed reaction, such that a 4:1 ratio of **10** and **9** was achieved through a thermal Diels–Alder reaction at 160°C.

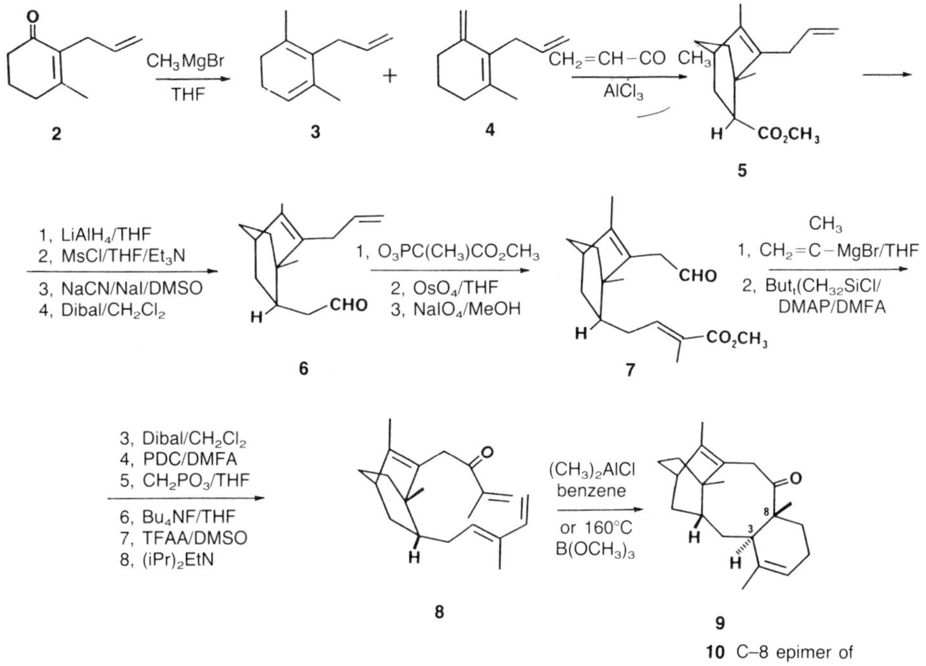

2 CH_3MgBr / THF → **3** + **4** $CH_2{=}CH{-}CO\ CH_3$ / $AlCl_3$ → **5** (CO_2CH_3)

1, $LiAlH_4$/THF
2, MsCl/THF/Et_3N
3, NaCN/NaI/DMSO
4, Dibal/CH_2Cl_2
→ **6** (CHO)

1, $O_3PC(CH_3)CO_2CH_3$
2, OsO_4/THF
3, $NaIO_4$/MeOH
→ **7** (CHO, CO_2CH_3)

1, $CH_2{=}C(CH_3){-}MgBr$/THF
2, $But_t(CH_3)_2SiCl$/DMAP/DMFA
3, Dibal/CH_2Cl_2
4, PDC/DMFA
5, CH_2PO_3/THF
6, Bu_4NF/THF
7, TFAA/DMSO
8, $(iPr)_2EtN$
→ **8**

$(CH_3)_2AlCl$ benzene or 160°C $B(OCH_3)_3$ → **9**

10 C–8 epimer of **9**

The synthetic hydrocarbon **18**, named *E,E*-verticillene, corresponds to verticillol (**19**) isolated from *Sciadopitys verticillata* Sieb. & Zucc. (Taxodiaceae) (Karlsson *et al.*, 1978). Due to the close skeletal similarity, it has been suggested that **18** is a precursor in the biosynthesis of the taxane group of alkaloids. An attempt has been made to synthesize *E,E*-verticillene (**18**) and to study the transannular cyclization of both **18** and **19** to the tricyclo[$10.3.1.0^{4,6}$]pentadecene carbon framework (the common skeleton of taxanes) in order to support the biogenetic pathway mentioned previously. *E,E*-Verticillene (**18**) has been obtained by the route shown, utilizing an intramolecular reductive coupling of bis-aldehyde **16** in the presence of Ti(0) with concomitant 1,5-H sigmatropic rearrangements, followed by 1,4-reduction using sodium in ammonia (Jackson and Pattenden, 1985). Treatment of **18** with Lewis acids, however, failed to produce the target taxane carbon framework (Begley *et al.*, 1985).

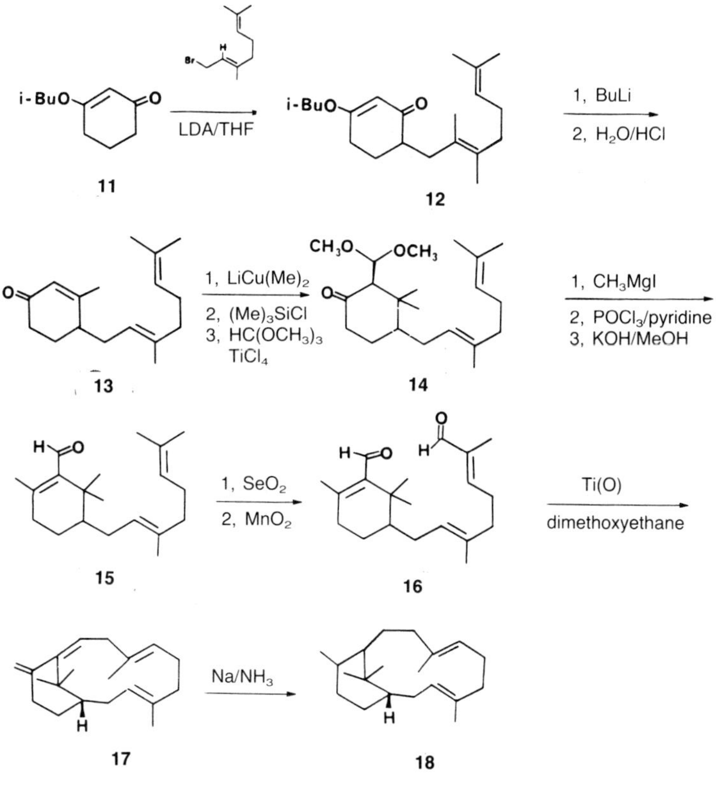

19

The first successful construction of the stereochemically correct, complete carbon framework of the taxanes was recently reported by Kende *et al.* (1986). A $TiCl_4$-mediated coupling of acetal **20** with enol silane **21** and subsequent acid treatment resulted in the enone diastereomers **22**. Selective vinyl cleavage gave the corresponding enone diesters **23**. Hydrogenation of the parent diastereomer enone followed by epimerization at C-3 led to the desired diester **24**. Methylenation of **24** followed by subsequent *iso*-$BuAlH_4$ reduction and Swern oxidation, afforded the dialdehyde **26** from which the sterically encumbered 8-membered B ring could be formed by McMurry cyclization. Product **27** is identical with the taxane skeleton in every stereochemical respect, moreover, it offers attractive potential for further functionalization in order to achieve natural taxol derivatives.

20 + 21 → ($TiCl_4/CH_2Cl_2$; p–TSA/benzene) → 22 → (OsO_4; $NaIO_4$)

→ (1, $NaClO_2$; 2, NH_2SO_3H/dioxane; 3, CH_2N_2) → 23 → (1, $H_2/Pd(C)$; 2, $K_2CO_3/MeOH$; 3, CH_2N_2)

→ 24 → ($ZN/CH_2Br_2/TiCl_4$; iBu_2AlH) → 25 → ($(CoCl)_2Me_2SO$)

→ 26 → ($TiCl_3$/Zn–Cu/DME) → 27

With the notion of producing compounds having enhanced biological activity, the chemical properties of taxol (**1**) have also been investigated (Magri and Kingston, 1986). Jones oxidation of **1** yielded 7-oxotaxol and 2′,7-dioxotaxol; however, their biological activities have not yet been reported. Several new taxol analogs were prepared by hemisynthesis starting from 10-deacetylbaccatin III, and their *in vitro* effect on the polymerization of tubulin was investigated (Senilh *et al.*, 1984). The presence of a hydroxyl group at C-2′ slightly improved the activity, while inversion of the groups at C-2′ and C-3′ decreased the activity, which was of the same order of magnitude as the activity of taxol. The action of taxol (**1**) on the cytoskeleton has been reviewed (Toyama, 1984).

b. Pyrrolizidine Alkaloids

Pyrrolizidine alkaloids exert remarkably diverse biological activities. On one hand, some of them exert antitumor, hypotensive, local anesthetic, antispasmotic, and anti-inflammatory activity, yet on the other hand several pyrrolizidine alkaloids are highly carcinogenic or hepatotoxic.

The chemistry and pharmacology of pyrrolizidine alkaloids have been reviewed in several different treatises (Robins, 1982, 1984, 1985; Suffness and Cordell, 1985; Wrobel, 1986), and detailed investigation of their biological activity has resulted in some structure–activity correlations. All of the active antitumor pyrrolizidine alkaloids contain an allylic alcohol function, e.g., supinine (**28**), heliotrine (**29**) and its *N*-oxide, crispatine (**30**), fulvine (**31**), monocrotaline (**32**), and indicine *N*-oxide (INO) (**33**). Indicine *N*-oxide (**33**) shows high-level activity against B16 melanoma, mammary xenograft, M5076 sarcoma, P388 leukemia, and Walker 256 carcinosarcoma. Indicine *N*-oxide has been submitted to Phase I and Phase II clinical trials, even though the dose is quite high.

Recently, a new type of pyrrolizidine alkaloid (**34**), with a unique 8-ethoxy-3-oxo-1,2-dehydroretrorsine skeleton, has been isolated from *Senecio grisbachii* Baker var. *grisbachii* (Hirschmann *et al.*, 1985). Dihydroretrorsine (**35**) has been obtained from the roots of *Senecio subulatus* Don ex Hook & Arn. var. *erectus* together with the previously known senecionine and retrorsine (Pestchanker *et al.*, 1985a). Reinvestigation of *Senecio latifolius* DC resulted in the isolation of a new pyrrolizidine alkaloid, mereskine *N*-oxide (**36**) (Bredenkamp *et al.*, 1985), while from the Western false forget-me-not, *Hackelia* *floribunda* (Lehm) Johnston, latifoline *N*-oxide (**37**) was isolated (Hagglund *et al.*, 1985). Uspallatine (**38**) has recently been obtained from the roots of *Senecio uspallatensis* (Pestchanker *et al.*, 1985b). X-Ray analysis of some pyrrolizidine alkaloids and derivatives have been reported (Mackay *et al.*, 1985; Wiedenfeld *et al.*, 1985). Negative ion mass spectra were recorded for senkirikine and fulvine (Madhusudanan *et al.*, 1984). Further methods for the determination of pyrrolizidine alkaloids and their metabolites have also been developed (Kedzierski and Buhler, 1986a).

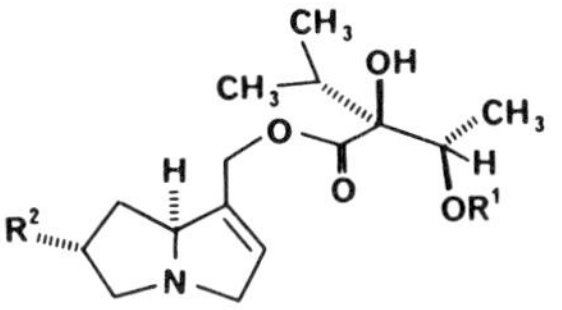

28 Supinine $R^1=H$, $R^2=H$

29 Heliotrine $R^1=CH_3$, $R^2=OH$

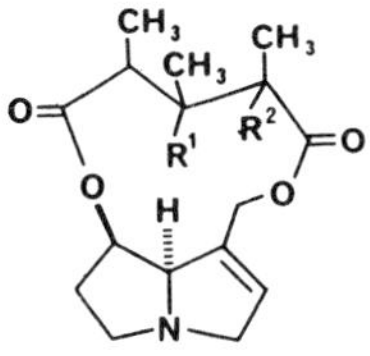

30 Crispatine $R^1=OH$, $R^2=H$

31 Fulvine $R^1=OH$, $R^2=H$

32 Monocrotaline $R^1=OH$, $R^2=OH$

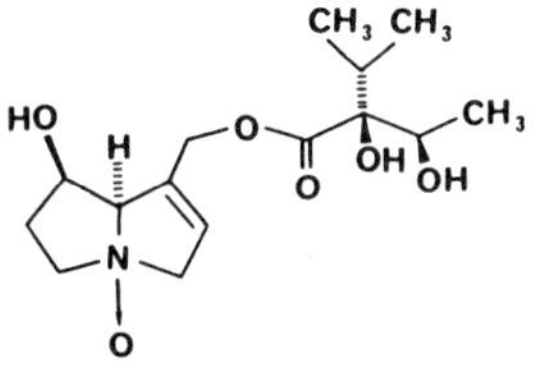

33 Indicine N-oxide

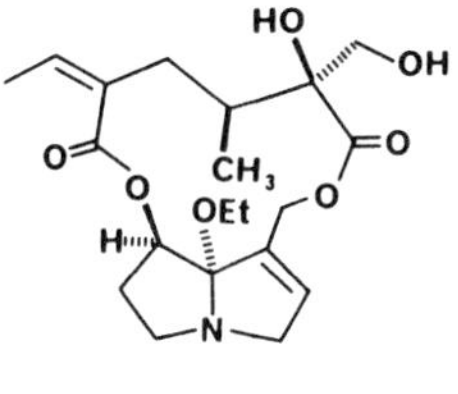

34

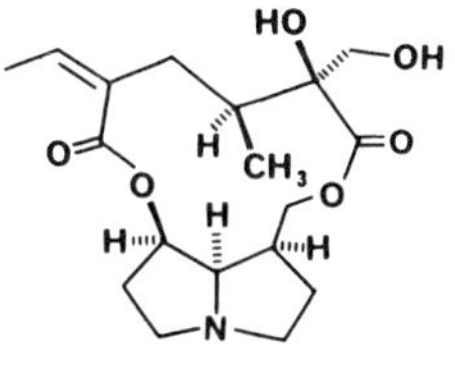

35 Dihydroretrorsine

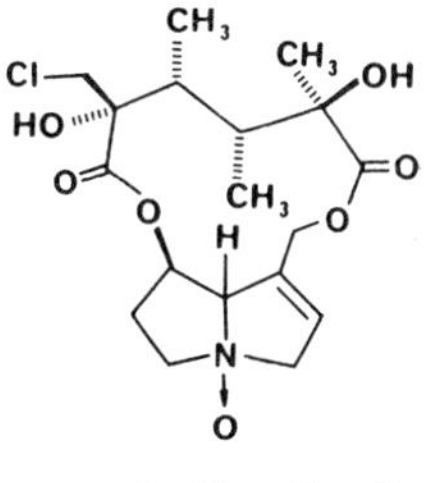

36 Mereskine N-oxide

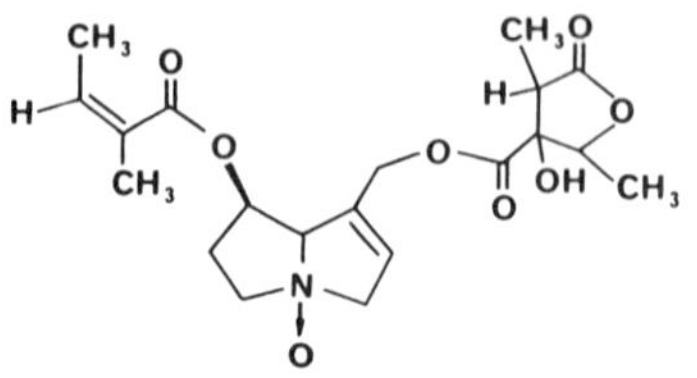

37 Latifoline N-oxide

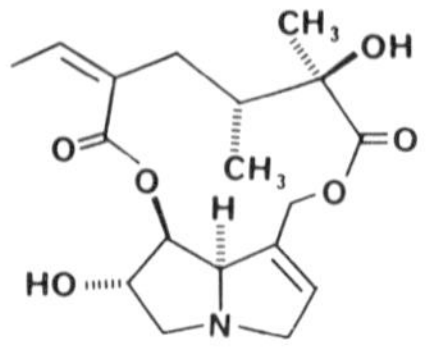

38 Uspallatine

The syntheses of the different necine bases and necine acids, and their coupling to open-chain or macrocyclic alkaloids have been thoroughly reviewed (Robins, 1984, 1985; Wrobel, 1986), consequently, only a brief summary will be presented here. A new technique, the utilization of nitrene cycloadditions, has been reported by several independent research groups for the construction of a pyrrolizidine ring system (Hudlicky *et al.*, 1985, 1986; Iida *et al.*, 1985; Lathbury and Gallagher, 1986; Pearson, 1985). Chamberlin and Chung (1983, 1985) accomplished the enantioselective total synthesis of seven pyrrolizidine diols starting from a single key-intermediate obtained from readily available (*S*)-malic acid. The syntheses of optically active pyrrolizidines utilizing a chiral starting material have also been reported (Ishibashi *et al.*, 1986; Nishimura *et al.*, 1985). The epoxidation of the diolefinic pyrrolizidine alkaloid seneciphylline (**39**) with performic acid has been studied in order to prepare senecicannabine (**40**), isolated recently from *Senecio cannabifolius* Less. The oxidation procedure resulted in two monoepoxides (**41** and **42**), as well as two diepoxides (**40** and **43**) whose steric structures were confirmed by [^{13}C]NMR analysis (Asada and Furuya, 1984).

The total synthesis of (−)-indicine *N*-oxide, the enantiomer of natural indicine *N*-oxide (**33**), has recently been achieved (Nishimura *et al.*, 1986). A series of papers has appeared on the total synthesis of optically active integerrimine, a 12-membered dilactonic pyrrolizidine alkaloid (Niwa *et al.*, 1986a,b,c).

The necic acids (−)- and (+)-trachelanthic acid, as well as (−)- and (+)-viridifloric acids, were synthesized from *trans*-α-isopropylcrotonic acid, and their isopropylidene derivatives were regiospecifically coupled with (−)-retronecine obtained from hydrolysis of the pyrrolizidine alkaloid monocrotaline. Hydrolysis, followed by oxidation, led to indicine *N*-oxide (**33**), intermedine *N*-oxide, and lycopsanine *N*-oxide (Zalkov *et al.*, 1985). Each of these alkaloids

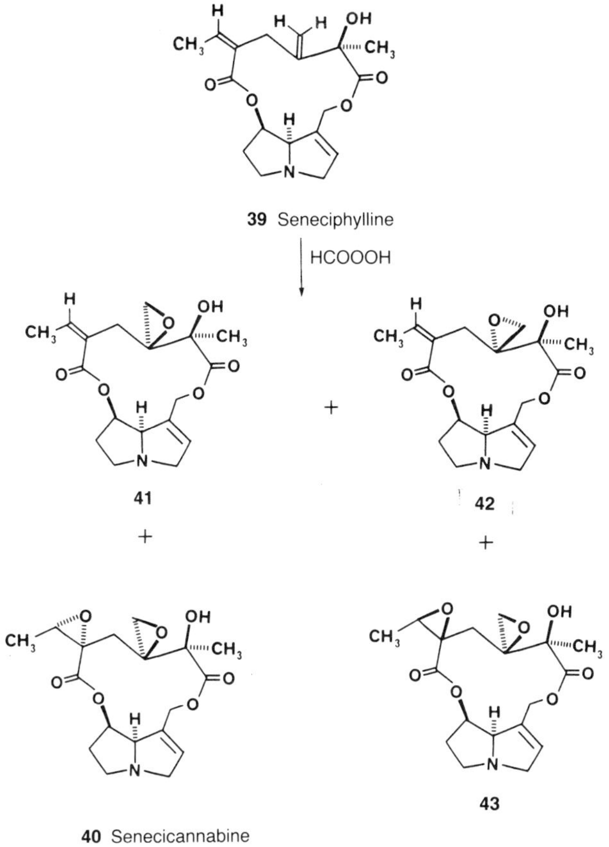

39 Seneciphylline

HCOOOH

41 + **42** +

40 Senecicannabine

43

and some of their derivatives was evaluated against the P388 lymphocytic leukemia. However, none of them proved to be more active than INO (**33**). Macrocyclic diesters of retronecine have been prepared, but no antitumor data have been reported (Burton and Robins, 1985). The genotoxicity of 17 pyrrolizidine alkaloids was tested in the hepatocyte primary culture–DNA repair test in order to examine their carcinogenicity, especially their ability to induce liver cancer (Mori *et al.*, 1985). Fifteen of these alkaloids demonstrated carcinogenic activity, i.e., monocrotaline, senecionine, seneciphylline, jacobine, epoxyseneciphylline, senecicannabine, senkirkine, fukinotoxin, acetylfukinotoxin, syneilesine, clivorine, dihydroclivonine, ligularidine, neoligularidine, and ligularizine. Only two alkaloids did not elicit DNA repair; retronecine, which lacks a necic acid moiety, and ligularizine, which is saturated at the 1,2-position of the pyrrolizidine ring. The mechanism by which pyrrolizidine alkaloids initiate and promote hepatic carcinogenesis (Hayes *et al.*, 1985), and also their toxic metabolites (Kedzierski and Buhler, 1985, 1986b; Lafranconi *et al.*, 1985; Segall *et al.*, 1985) have also been studied.

c. Acronycine

Acronycine (**44**), an acridone alkaloid first isolated from the bark of *Acronychia baueri* Schott. (Rutaceae), exerts a broad spectrum of *in vivo* antineoplastic activity. Chemical properties, synthesis, as well as biological activity of acronycine (**44**) and its derivatives have recently been reviewed (Gerzon and Svoboda, 1983; Suffness and Cordell, 1985). It is worth mentioning that acronycine (**44**) neither interacts with DNA nor affects RNA synthesis; two of the most common mechanisms of action of antitumor agents. Further work on the mechanism of action is currently under way in our laboratories.

As a continuation of the study (Funayama *et al.*, 1983, 1984, 1985; Funayama and Cordell, 1983, 1984, 1985a) of the chemistry of acronycine, Cordell and co-workers have reported on several new pentamers from the acid treatment of noracronycine (Funayama and Cordell, 1986) and on the mechanism of the rearrangement of the angular to the linear skeleton (Funayama and Cordell, 1985b). In order to potentiate the availability of functional groups thought to be important for anticancer activity, the synthesis of acronycine dimers through C-7 was developed (Gunawardana and Cordell, 1987). Two types of dimers were prepared. Treatment of thioacronycine with diazomethane resulted in a dimer in which the two acronycine units were linked through an unstable cyclopropane ring. The other types of dimers consist of isomers of azine (**45**), obtained from acronycine (**44**) through chlorination and condensation with hydrazine. Biological evaluation of these derivatives is presently under way.

The aza analog of acronycine was prepared; however, compound **46** proved to be inactive in the tumor leukemia tests screened (Reisch and Aly, 1986).

d. *Cephalotaxus* Alkaloids

The active antitumor principles of *Cephalotaxus harringtonia* var. *drupacea* are harringtonine (**48**) and homoharringtonine (**49**), and each has been selected for preclinical development by the NCI.

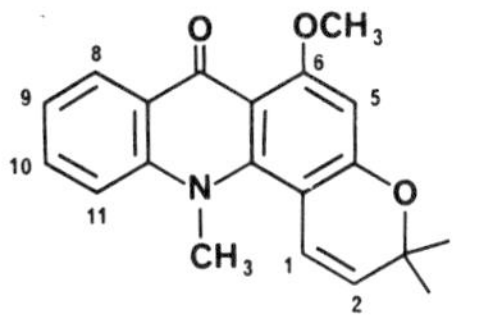

44 Acronycine

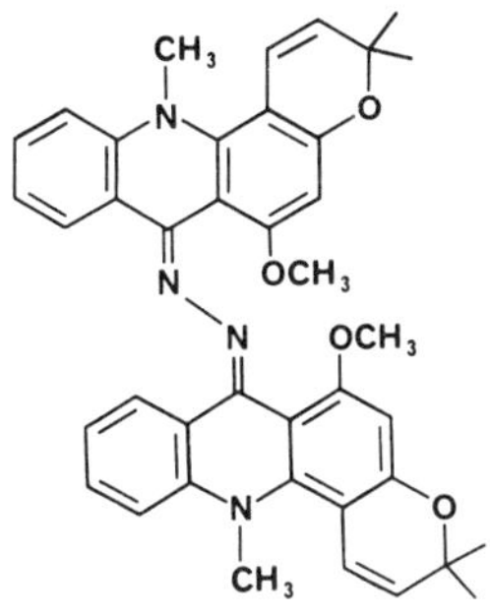

45

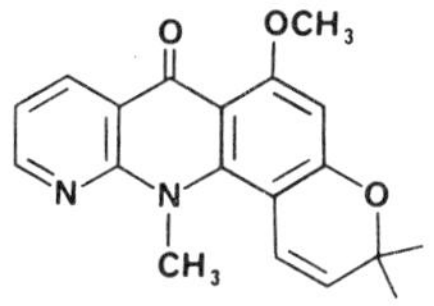

46

Several excellent recent reviews (Findlay, 1976; Huang and Xue, 1984; Smith *et al.*, 1980; Suffness and Cordell, 1985; Weinreb and Semmelhack, 1975) on the isolation, preparation, and pharmacological activity of *Cephalotaxus* alkaloids are available, so discussion of these aspects is quite limited here.

A new minor alkaloid, 2-epicephalofortuneine, has been isolated from *Cephalotaxus fortunei* Hook. The structure of this new alkaloid was postulated to be **52** by spectroscopic analysis; however, the structure elucidation is still incomplete (Lin *et al.*, 1985).

In order to accomplish the synthesis of the *Cephalotaxus* alkaloids, two major preparative problems have to be solved. On one hand, there is the synthesis of the cephalotaxine nucleus (**47**) and the esterifying acid, and on the other hand, there is the coupling of these

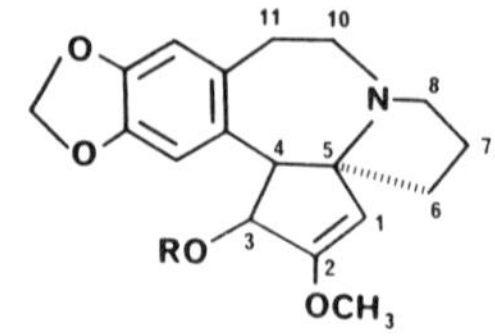

47 Cephalotaxine R=H

48 Harringtonine

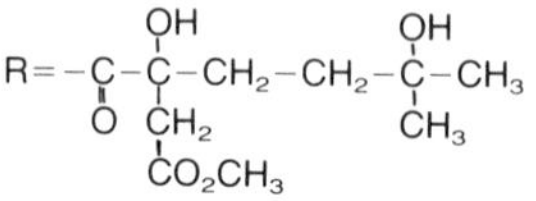

49 Homoharringtonine

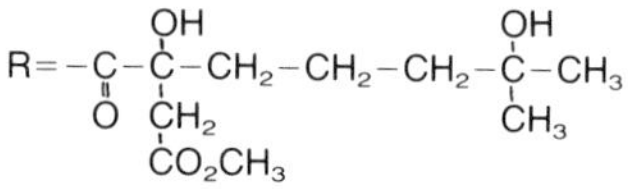

50 Isoharringtonine

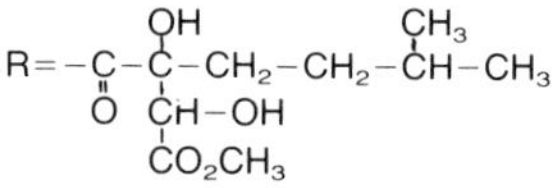

51 Deoxyharringtonine

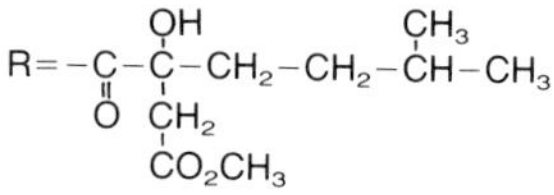

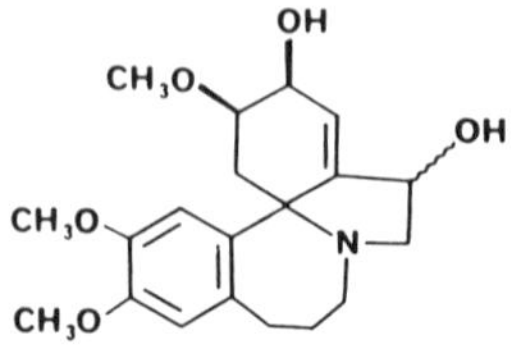

52

two units through an esterification process. Several attempts have been reported (Findlay, 1976; Huang and Xue, 1984) for the synthesis of the tetracyclic cephalotaxine skeleton; however, the stereoselective coupling is still a challenging problem. Several groups in the People's Republic of China are studying the latter problem (Huang and Xue, 1984), and the synthesis of homoharringtonine (**49**) and isoharringtonine (**50**) were recently reported. The Grignard reaction of 1-bromo-4-(ethylenedioxy)pentane with diethyl oxalate gave ethyl 2-oxo-6-(ethylenedioxy)heptanoate, which was converted to the corresponding cephalotaxine ester **53**. The Reformatsky reaction of **53** with ethyl bromoacetate, followed by deketalization and a Grignard reaction with methylmagnesium iodide, afforded homoharringtonine (**49**) and its epimers (Wang *et al.*, 1985). Isoharringtonine (**50**) and three of its stereoisomers were prepared by a similar route. The Reformatsky reaction of the appropriate cephalotaxine ester **54** with methyl(benzyloxy)bromoacetate, and subsequent hydrogenolysis, resulted in four diastereomers of isoharringtonine (**50**), which were separated by a Chromatotron (Li *et al.*, 1984a). Isoharringtonine (**50**) has also been synthesized by treating **54** with methyl glyoxylate in the presence of cyclopentadienyltrichlorotitanium and lithium aluminum hydride (Li *et al.*, 1984b).

Stereospecific synthesis of deoxyharringtonine (**51**) has been performed by condensation of **54** with (±)-(*R*)-4-CH_3-C_6H_4SO-$CH_2CO_2C(CH_3)_3$ followed by desulfurization, hydrolysis, and methylation (Cheng *et al.*, 1984).

53 → 49: 1, $BrCH_2CO_2CH_3/Zn$; 2, H^+; 3, CH_3MgI

54 → 50: 1, $Br-CH(OCH_2O)-CO_2CH_3$; 2, $H_2/Pd(C)$

54 → 51: 1, (p-tolyl sulfinyl reagent); 2, desulfurization; 3, hydrolysis; 4, esterification

Both harringtonine (**48**) and homoharringtonine (**49**) are effective against P388 lymphocytic leukemia, as well as against L1210 leukemia. *Cephalotaxus* alkaloids generally inhibit the biosynthesis of DNA. Studies on the kinetics of the competitive inhibition of DNA polymerase have established that the inhibitory effect of homoharringtonine (**49**) is stronger than that of harringtonine (**48**) or isoharringtonine (**50**) (Qiu and Wu, 1984). *Cephalotaxus* alkaloids also inhibit protein synthesis in HeLa cells. According to studies on the mechanism of this action, harringtonine (**48**) seems to inhibit protein synthesis at a step before the formation of the first peptide bond (Wu *et al.*, 1984). Recent progress in the studies of the antitumor activity of harringtonine (**48**) and homoharringtonine (**49**) have been summarized (Kato *et al.*, 1984; Li and Han, 1986).

e. Colchicine

The chemistry and biological properties of colchicine (**55**) and its derivatives have recently been reviewed in detail by Capraro and Brossi (1984), including the relationship between structure, toxicity, and antitumor activity. Colchicine (**55**), the major alkaloid of *Colchicum autumnale* L., and related compounds generally exert antimitotic properties, interfere with microtubule-dependent cell functions, and irreversibly bind to tubulin.

Speciosamine (**58**) has recently been reported (Chommadov *et al.*, 1985) as a new base from *Colchicum speciosum* Stev. The biosynthesis of colchicines, their effect on microtubular systems of cells, and the use of colchicine as an antichemotaxis agent have been summarized (Thies, 1985).

Because colchicine (**55**) itself is too toxic for human use as an antitumor drug, substantial effort has been made to obtain colchicine derivatives of reduced toxicity. Demecolcine (**56**, DMC) and trimethylcolchicine acid methyl ester (**57**, TMCA) are considerably less toxic than **55** and were selected for clinical evaluation. Demecolcine (**56**) is effective against myelocytic leukemia, while TMCA (**57**) might be useful in the treatment of chronic granulocytic leukemia.

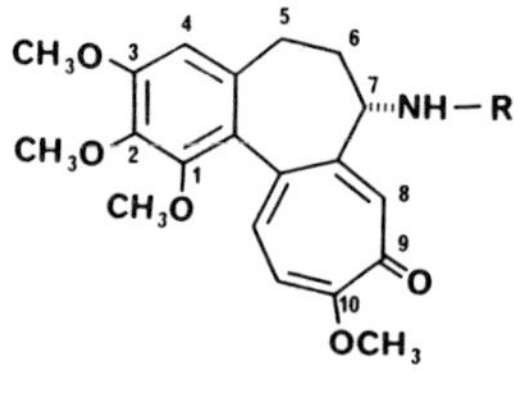

55 Colchicine $R=COCH_3$
56 DMC $R=CH_3$
57 TMCA $R=H$

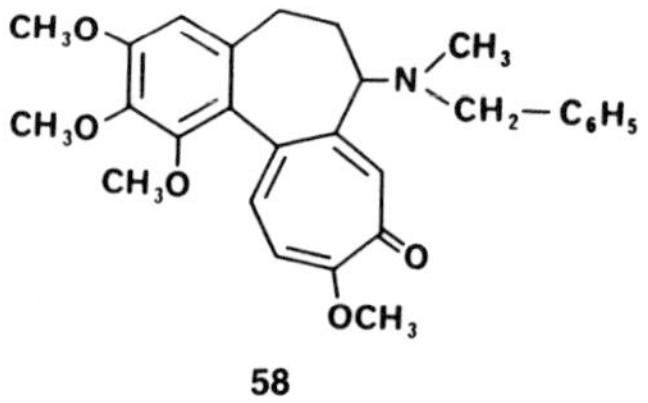

58

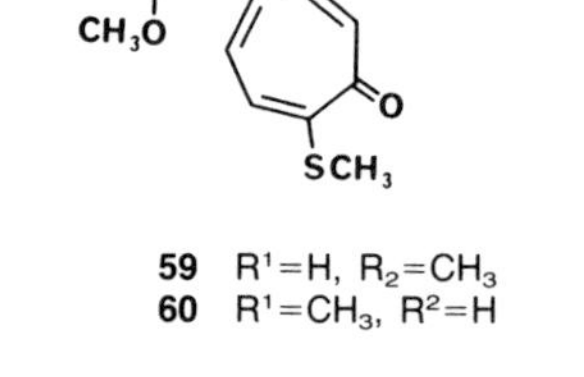

59 $R^1=H$, $R_2=CH_3$
60 $R^1=CH_3$, $R^2=H$

Through systematic variation of the substitution of ring A and the acetamido group of ring B of colchicine, Brossi and co-workers developed a new series of highly active antileukemic colchicine derivatives, e.g., 2-demethyl- and 3-demethylthiocolchicines (**59** and **60**, resp.) According to the biological data collected, **60** represents a broad-spectrum antitumor agent of some promise (Kerekes *et al.*, 1985a). The carbamates of colchicine and thiocolchicine proved to be suitable agents for the treatment of gout and murine malignancies (Brossi and Kerekes, 1984). Ester derivatives of 1-*O*-demethylthiocolchicine have also been prepared (Kerekes *et al.*, 1985b).

A study of the relationships between chemical structure and antimitotic activities of the colchicines showed that enhanced activity can be expected in those compounds containing an amide function at C-7. Replacement of the nitrogen by an alkyl group results in a decrease in activity (Dvorackova *et al.*, 1985). The effect on the binding of colchicine (**55**) to tubulin from beef brain was investigated in order to establish its potential as a model for new antimicrotubular drugs to bind to the colchicine binding site (Luyckx *et al.*, 1984). Colchicine (**55**) itself acts as a substoichiometric microtubular poison (Saxton *et al.*, 1984), and the neurotoxicity of colchicine (Steward *et al.*, 1984) as well as its effects on the cytoskeleton and mutagenic activity (Ishikawa, 1984; Balkandzhieva, 1985) have been reviewed.

f. Camptothecine

Camptothecine (**61**) and related alkaloids are a well-established series of antitumor agents, and camptothecine sodium has been subjected to Phase I clinical trials in both the United States and the People's Republic of China. 10-Hydroxycamptothecine (**62**), originally isolated from *Camptotheca acuminata* DC., and having a better therapeutic index than camptothecine (**61**), is presently under Phase II clinical study. It has been established that 10-hydroxycamptothecine (**62**) is a potent anticancer agent against several malignancies in man, and it is widely used in China for the treatment of liver cancer, leukemia, and gastric cancer. Recent results on its development as an antitumor agent have been summarized (Xu and Yang, 1984, 1985) and some of its derivatives have also been investigated (Yakult Honsha Co., 1985).

The synthesis of camptothecine and related compounds has attracted wide attention in the past 20 years, and these efforts together with the pharmacological data have been extensively reviewed (Cai and Hutchinson, 1983a,b; Hutchinson, 1981; Schultz,

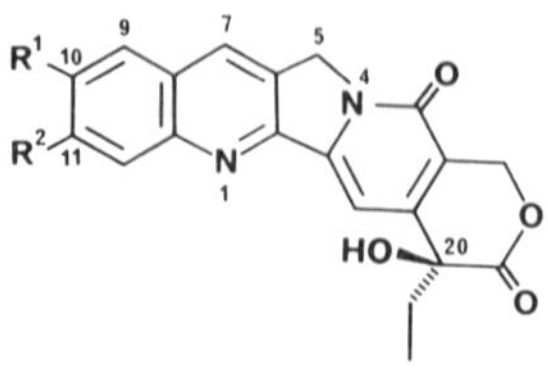

61 Camptothecine $R^1=H$, $R^2=H$

62 10–Hydroxycamptothecine $R^1=OH$, $R^2=H$

63 $R^1=OCH_3$, $R_2=H$

64 $R^1=R_2=OCH_3$

1973; Shamma and Georgiev, 1974; Wall and Wani, 1980; Winterfeldt, 1975).

Structure–activity relationship studies on camptothecine (**61**) and related compounds demonstrated that the α-hydroxy-lactone moiety and the conjugated linkage in rings A, B, C, and D are requisite for the antineoplastic activity. Interestingly, whereas 10-hydroxycamptothecine (**62**) is highly active, the 10-methoxy analog (**63**) shows only moderate activity and 10,11-dimethoxycamptothecine (**64**) is inactive (Wall and Wani, 1984). In combination with copper(II) and UV light, camptothecine (**61**) produced remarkable DNA strand scission (Kuwahara *et al.*, 1985). In addition to potent single-strand DNA breaking properties, camptothecine (**61**) also acts as a strong inhibitor of nucleic acid synthesis in mammalian cells (Hsiang *et al.*, 1985). The mechanism of action of camptothecine (**61**) and its derivatives on DNA has also been investigated (Fukada, 1985) and a theoretical model for characterizing camptothecine-binding to plasma proteins has been proposed (Boxenbaum and Fertig, 1984).

g. Ellipticine

The ellipticines comprise one of the most important groups of antitumor alkaloids. Ellipticine (**65**) and 10-methoxyellipticine[1] (**66**) exert marked activity against the L1210, P388, and P1534 leukemias, the X5563 myeloma, and Gardner lymphosarcoma. The principal mechanism of their antitumor activity is related to their planar structure and to their ability to intercalate with DNA base pairs. The broad spectrum of activity of ellipticine (**65**), 10-

[1] There are two numbering systems used for ellipticine derivatives. Throughout the chapter we have used the numbering system based on the biogenetic pathway.

methoxyellipticine (**66**), and the high activity of some synthetic derivatives, e.g., 4-methyl-10-hydroxyellipticinium acetate (**67**) and the 10-aza analogs **68** and **69**, make this group of compounds extremely promising for further development. The synthesis of ellipticines and related derivatives up to 1985 has been reviewed (Cordell, 1979; Gribble, 1985; Hewlins *et al.*, 1984; Sainsbury, 1977; Suffness and Cordell, 1985) and their pharmacological properties have also been summarized (Suffness and Cordell, 1985; Suffness and Douros, 1980).

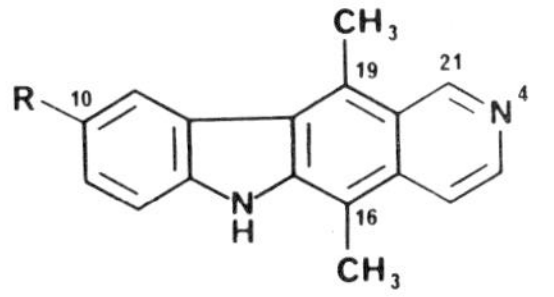

65 Ellipticine R=H

66 10–Methoxyellipticine R=OCH_3

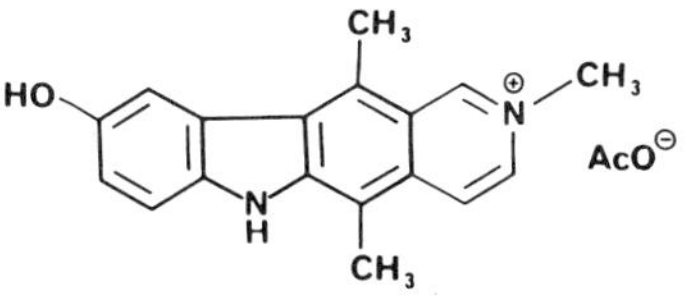

67

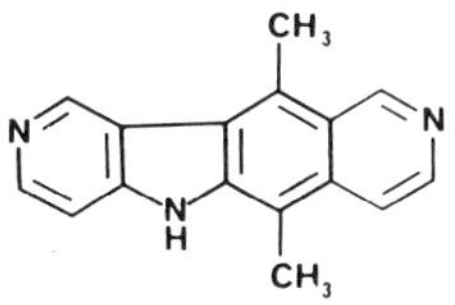

68

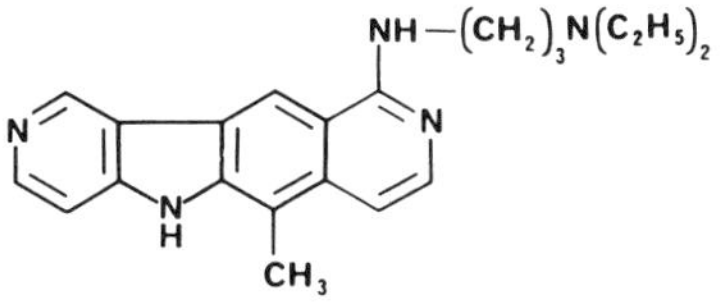

69

Recently, a facile synthesis of *N*-methyl-tetrahydroellipticine (**70**) has been reported (Differding and Ghosez, 1985) starting from *N*-methyl-piperidone (**72**) *via* the intramolecular Diels–Alder cycloaddition of acetylenic vinylketeninine (**75**) generated from β,γ-unsaturated anilide (**73**) with triphenylphosphine and bromine in the presence of triethylamine. Cycloadduct **76** was readily transformed into **70** by lithium aluminum hydride reduction.

A formal total synthesis of ellipticine (**65**) has been achieved (Ketcha and Gribble, 1985) by Friedel-Crafts acylation of 1-(phenylsulfonyl)indole (**77**) with 4-carbomethoxynicotinoyl chloride (**78**) followed by base-induced cyclization of **79** to produce ellipticine quinone (**80**). The latter was previously converted into ellipticine by simple reaction steps (Gribble, 1985).

The synthesis and antitumor activity of 10-hydroxyolivacine (**81**) and 12-hydroxy-olivacine (**82**) have been reported (Maftouh *et al.*,

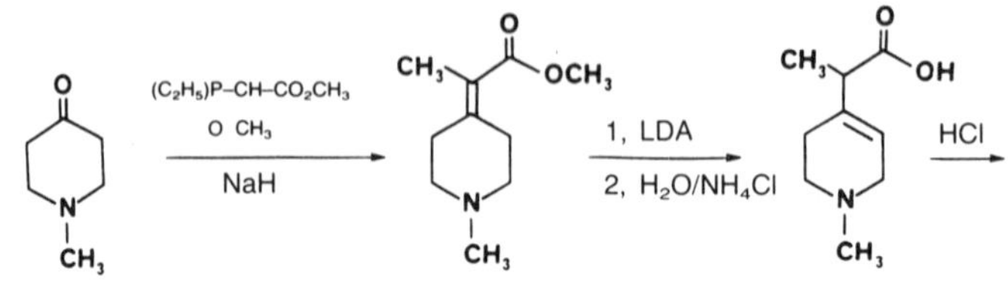

71 72 73

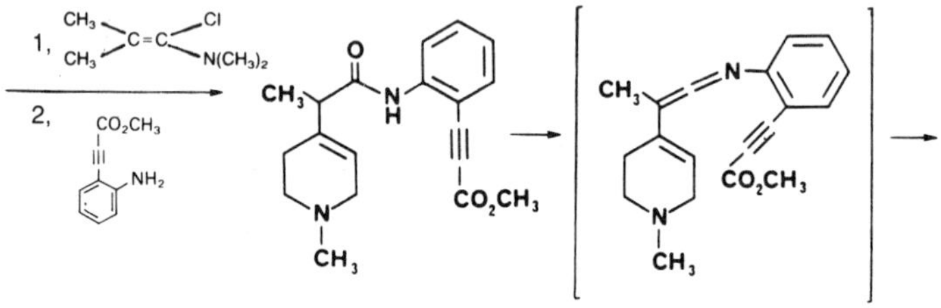

74 75

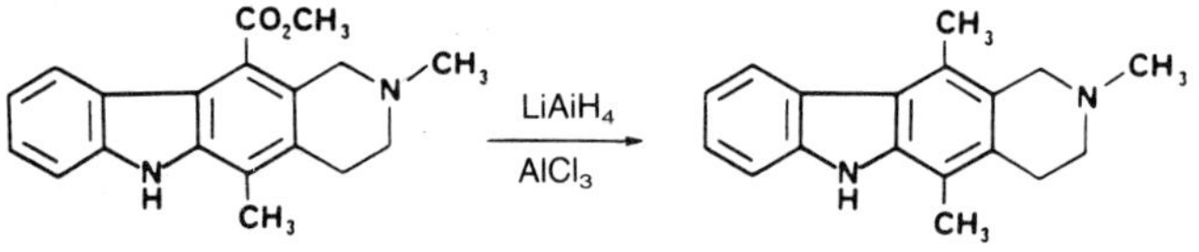

76 70

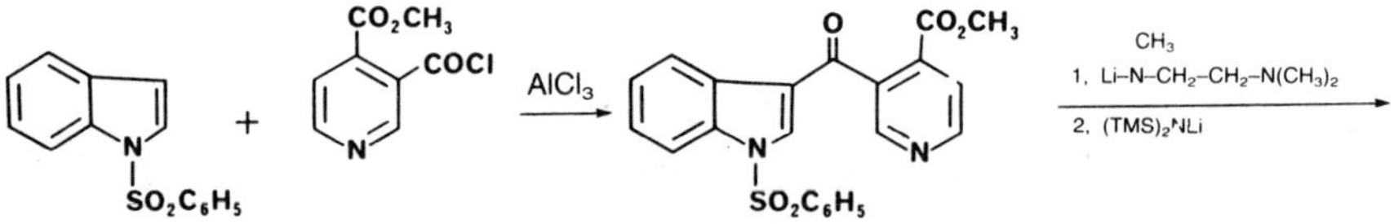

77 78 79

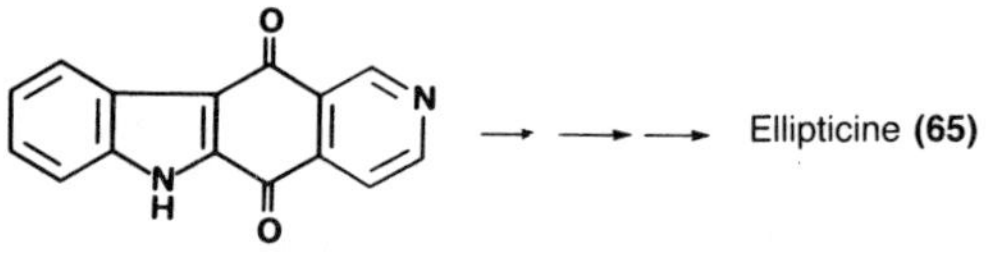

80

1985a). 5-Methoxyindole (**84**) and 7-methoxyindole (**85**) were condensed with the Δ^3-piperidine derivative **86**. Compounds **87** and **88** were acetylated to the corresponding amides, which were then cyclized via the Bischler–Napieralski route and reduced to tetrahydropyrido[4,3-*b*] carbazoles **89** and **90**, respectively. Demethylation and dehydrogenation, followed by cleavage of the methoxyl groups at C-10 and C-12 with hydrogen bromide, gave 10- and 12-hydroxyolivacines (**81** and **82**), respectively. The two hydroxyolivacines (**81** and **82**) have been identified as *in vivo* metabolites of olivacine (**83**) in the rat. Hydroxylation at position 10 enhanced the *in vitro* cytotoxicity against L1210 leukemia cells (ID_{50} = 0.06 μM compard to 2.03 μM for olivacine). However, the opposite effect was observed for 12-hydroxyolivacine (ID_{50} = 12.8 μM) (Maftouh *et al.*, 1985a).

AcOH

84 R^1=OH, R^2=H
85 R^1=H, R^2=OH

86

87 R^1=OH, R^2=H
88 R^1=H, R^2=OH

1, Ac_2O
2, $POCl_3$
3, $NaBH_4$

89 R^1=OH, R^2=H
90 R^1=H, R^2=OH

1, Pd(C)/decalin
2, HBr

81 10–Hydroxyolivacine R^1=OH, R^2=H
82 12–Hydroxyolivacine R^1=H, R^2=OH
83 Olivacine R^1=R^2=H

N^2-Methyl-10-hydroxyellipticinum (**91**) exhibits very good cytotoxic activity against a variety of tumor cells and is used in the treatment of breast cancer metastases. In order to improve the membrane transport of **91**, derivatives were prepared with improved lipophilicity (Auclair *et al.*, 1984). N^2-Methyl-10-hydroxyellipticinum (**91**) was readily oxidized with peroxidase–hydrogen peroxide to the corresponding quinone imine (**92**). The mesomeric form of **92** can be attacked by a range of nucleophiles such as the amino group of α-amino acids, e.g., glycine, alanine, valine, or leucine. Rearomatization of the intermediate adduct, followed by spontaneous oxidation, led to compounds **93–96**, respectively. All of the prepared adducts exhibited higher lipophilic character than did **91**. However, the amino acid moiety linked to the pyrido [4,3-*b*]carbazole ring system resulted in a slight decrease in cytotoxicity on L1210 cells *in vitro* compared with **91**. The coupling reactions of **91** with more complex nucleophiles have also been studied (Auclair *et al.*, 1983, 1986; Gouyette *et al.*, 1985; Pratviel *et al.*, 1985a,b).

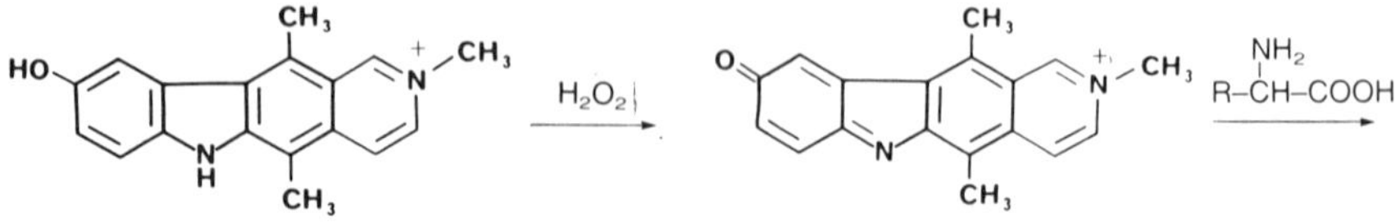

91

H_2O_2

92

NH_2
R–CH–COOH

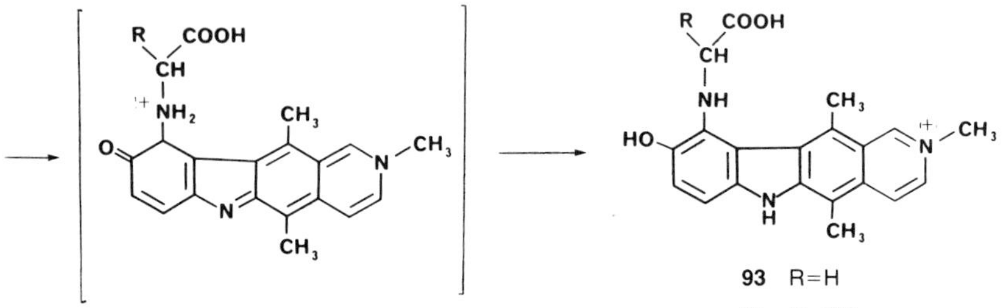

93 R=H

94 $R=CH_3$

95 $R=CH(CH_3)_2$

96 $R=CH_2CH(CH_3)_2$

Aminomethyl-substituted olivacine derivatives such as **97** have shown only weak antitumor activity (Jatztold-Howorko *et al.*, 1984). Compound **98**, a linked derivative of ellipticine to estradiol, retains the antitumor efficacy of ellipticine (**65**). Competition experiments with estradiol on the hormone-dependent human MCF-7 breast cancer cell line demonstrated that transport by the estrogen receptor system is not involved in the antitumor activity of **98** (Delbarre *et al.*, 1985).

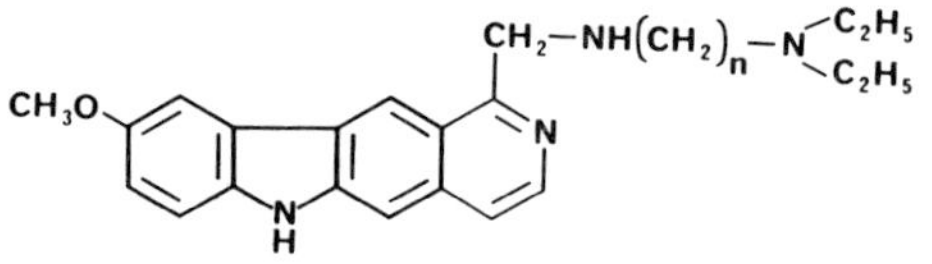

97 n=2 or 3

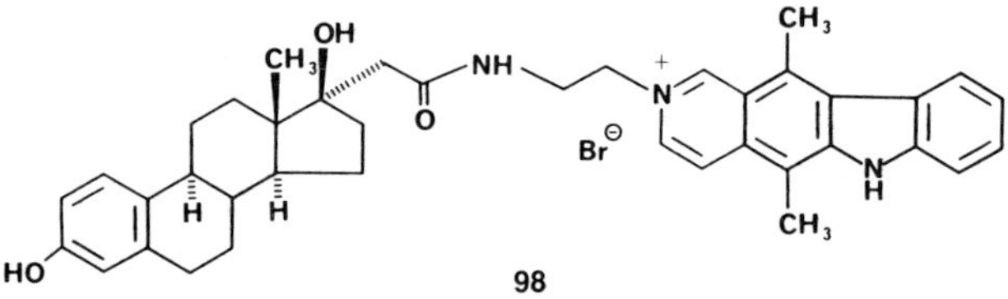

98

The effect of membrane potential on the cellular uptake of N^2-methyl-ellipticinium has been investigated (Charcosset *et al.*, 1984), and the teratogenic effects of 10-hydroxyellipticine in the mouse have been reported (Cros and Raynaud, 1984). Details of the metabolism of N^2-methyl-10-hydroxyellipticinium (**91**) have been published (Maftouh *et al.*, 1985b), and the scission of DNA apurinic sites by 10-aminoellipticine has been demonstrated (Malvy *et al.*, 1986). A correlation has been observed between the *in vitro* DNA intercalation effect of ellipticines and their physiological properties (Bertrand and Giacomoni, 1985). The ability of ellipticines to intercalate with DNA has also been studied using molecular graphics (Kuroda and Sainsbury, 1984).

h. Emetine

Emetine (**99**), the major alkaloid of *Cephaelis ipecacuanha* Rich., and its 2,3-dehydro derivative are used clinically in the treatment of amebiosis. Recent investigations have established that emetine (**99**) is effective against L1210 as well as P388 leukemias, and thus the emetine alkaloids have become a target of interest as potential antitumor agents (Suffness and Cordell, 1985). Complete listings of emetine alkaloids and their natural sources have appeared (Wiegrebe *et al.*, 1984). (−)-9-Demethyltubulosine (**100**), a new alkaloid from *Alangium vitiense* (A. Gray) Baillon, has recently been isolated and its structure established spectroscopically (Kan-Fan *et al.*, 1985).

The synthesis of emetine alkaloids and related derivatives has been reviewed (Brossi *et al.*, 1971; Fujii and Ohba, 1983; Shamma, 1972a; Shamma and Moniot, 1978a). Fujii and Ohba (1985a) recently described the stereospecific synthesis of (±)-9-demethylpsy-

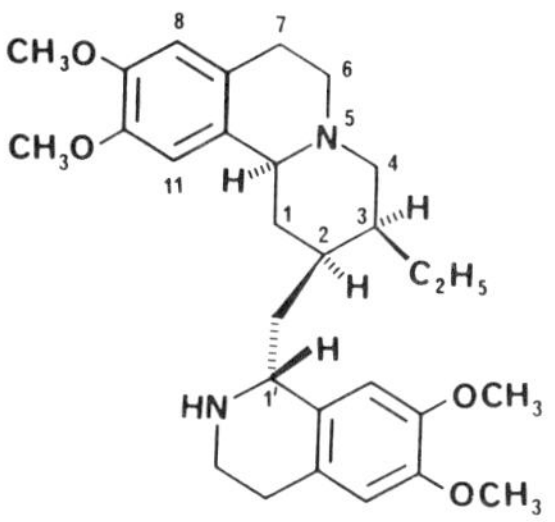

99 Emetine

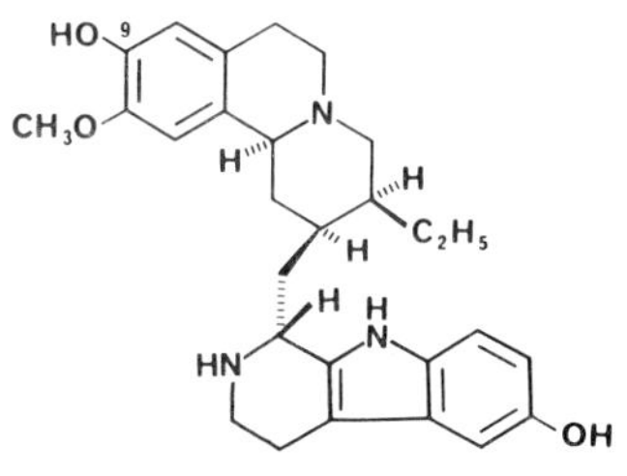

100 9-Demethyltubulosine

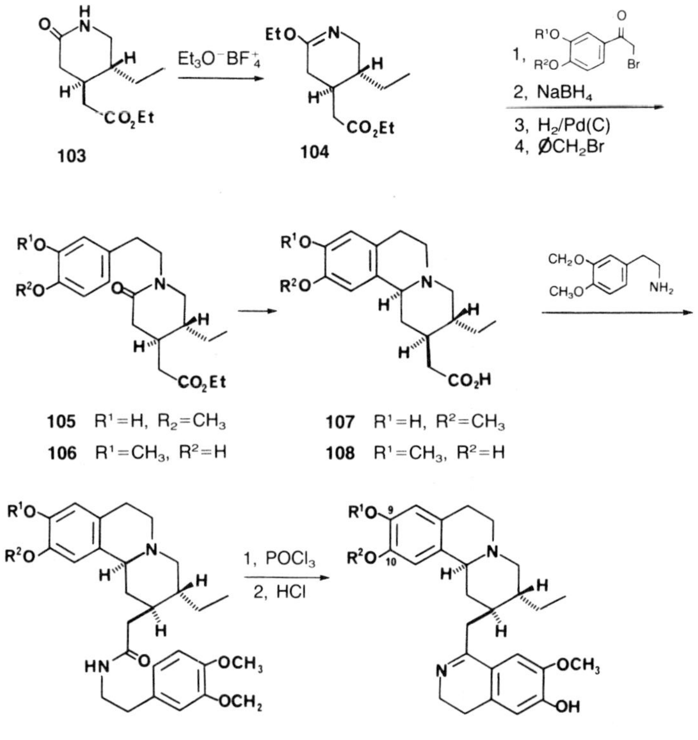

chotrine (**101**) and (±)-10-demethylpsychotrine (**102**), with a view to the structure elucidation of "desmethylpsychotrine" isolated previously (Pakrashi and Ali, 1967) from *Alangium lamarckii* Thwaites. The synthesis began with the condensation of lactim ether (**104**) with a phenacyl bromide derivative, followed by simple transformation steps into key-intermediates **107** and **108**, respectively. The formation of the second isoquinoline moiety was performed by condensation with 3-benzyloxy-4-methoxyphenethylamine and subsequent Bischler–Napieralski cyclization. Deprotection of **109** and **110** afforded 9-demethylpsychotrine (**101**) and 10-demethylpsychotrine (**102**), respectively. Spectral comparison of (±)-**101** and (±)-**102** with natural "desmethylpsychotrine" gave evidence that formula **101** represents the structure of the natural alkaloid. The same compound, (+)-**101**, has also been synthesized (Fujii and Ohba, 1985b) from optically active **103**, prepared from cinchonine by a classical degradation procedure utilizing the lactim ether method described above. The identity of synthetic (+)-**101** with natural "desmethylpsychotrine" unequivocally established the absolute configurations of stereo centers at C-2, C-3, and C-11b.

The same research group has also reported the first total synthesis of (±)- and (−)-alangimarckine (**111**)(Fujii *et al.*, 1985). Utilizing their lactim ether method for the construction of rings A, B, and C, intermediate **112** was prepared. Condensation of **112** with tryptamine, followed by subsequent Bischler–Napieralski cyclization and catalytic hydrogenation, led to alangimarckine (**111**). The reaction sequence was carried out on both the racemic and optically active series of intermediates. Identity of synthetic (−)-**111** with natural alangimarckine unequivocally established the structure and absolute configuration of this alkaloid.

A series of ochrolifuanine derivatives (**114**), a group of bisindole alkaloids similar to emetine, were prepared by Pictet–Spengler condensation of dihydrocorynantheal (**113**) with variously substituted tryptamines. Compounds of type **114** were cytotoxic to proliferating meristem cells of *Allium sativum* L. *in vitro*; however, they were inactive against P388 leukemia in mice (Seguin *et al.*, 1985).

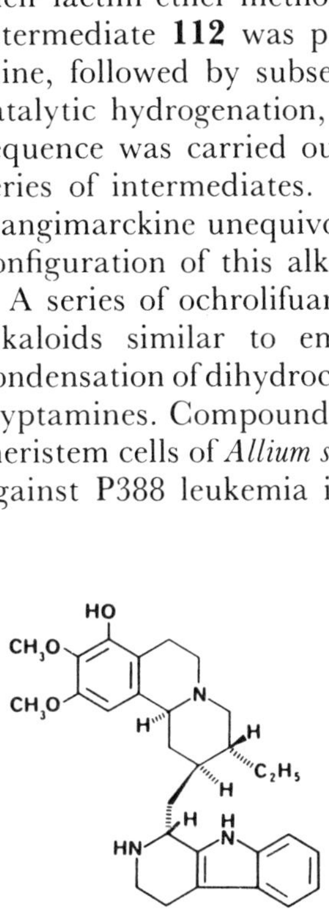

111

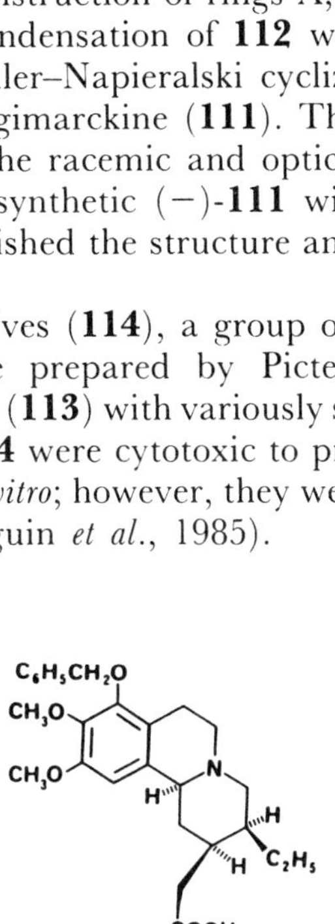

112

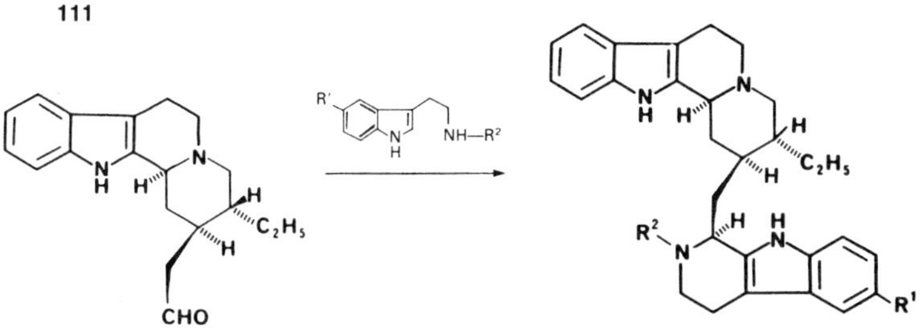

113

114 R^1=H, OH, CH_3 or OCH_3
R_2=H or CH_3

i. Phenanthroindolizidine and Phenanthroquinolizidine Alkaloids

Phenanthroindolizidine, as well as phenanthroquinolizidine, alkaloids are relatively small groups of alkaloids, each biogenetically formed from phenylalanine and tyrosine. The antitumor activities of some of the representatives of these alkaloids, e.g., tylocrebrine (**115**), tylophorine (**116**), and cryptopleurine (**117**), are weaker than those of emetine (**99**), and since the mechanism of action of compounds **115–117** is very similar, further development as antitumor agents may be anticipated as in the emetine series.

A new phenanthroindolizidine alkaloid, tylophorinicine (**118**), has been isolated from the roots of *Tylophora asthmatica* Wight & Arn. and *Pergularia pallida* Wight & Arn. (Mulchandani and Venkatachalam, 1984). Four new alkaloids, tylohirsutinine (**119**), tylohirsutinidine (**120**), 13*a*-methyltylohirsutine (**121**) and 13*a*-methyltylohirsutinidine (**122**), as well as a diarylindolizidine derivative, 13*a*-hydroxysepticine (**123**), were obtained from *Tylophora hirsuta* Wight (Bhutani *et al.*, 1984). Besides the known alkaloid cryptopleurine (**117**), a new diarylquinolizidine, kayawongine (**124**), has been isolated from *Cissus rheifolia* Planch (Saifah *et al.*, 1983). The pharmacological properties of compounds **118–124** have not been reported.

Developments on the occurrence and synthesis of phenanthroindolizines and phenanthroquinolizidine alkaloids have been reviewed (Lamberton, 1984; Grundon, 1985a). Recently, Iida *et al.* (1984) described a general synthetic route to obtain both phenanthroindolizidine and phenanthroquinolizidine alkaloids. Dipolar cycloaddition of either pyrroline *N*-oxide or tetrahydropyridine *N*-oxide with 3,4-dimethoxystyrene (**125**) gave a mixture of diastereomeric adducts

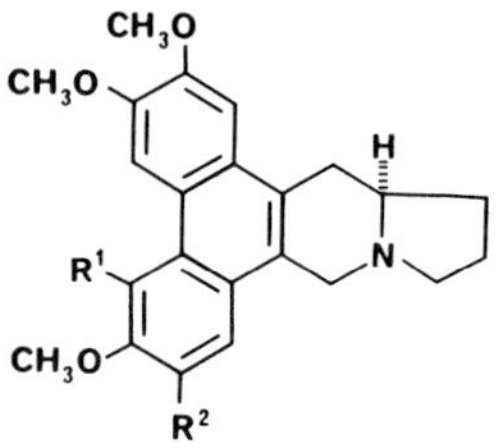

CH_3O CH_3O H N R^1 CH_3O R^2

115 Tylocrebrine R^1=OCH_3, R^2=H

116 Tylophorine R^1=H, R^2=OCH_3

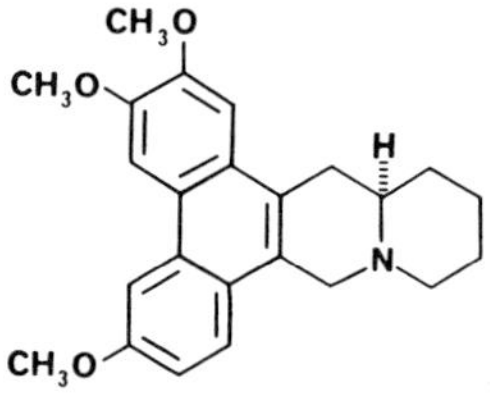

CH_3O CH_3O H N CH_3O

117 Cryptopleurine

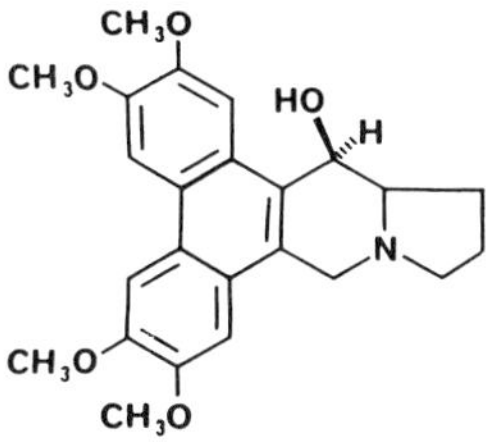

118 Thlophorinicine

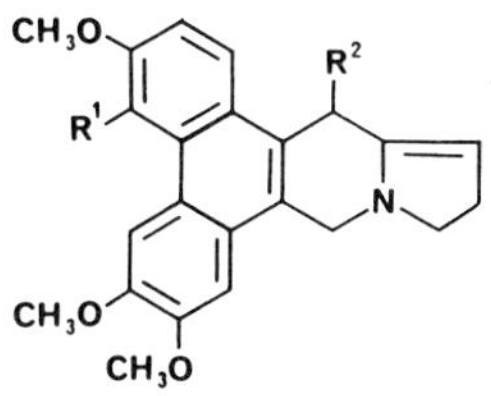

119 Tylohirsutinine
$R^1=OCH_3$, $R^2=H$

120 Tylohirsutinidine
$R^1=R^2=OH$

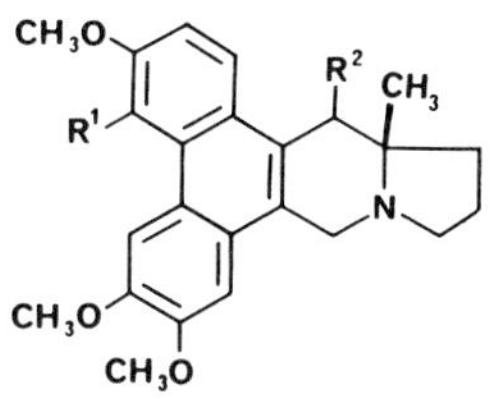

121 13a-Methyltylohirsutine
$R^1=OCH_3$, $R^2=H$

122 13a-Methyltylohirsutinidine
$R^1=R^2=OH$

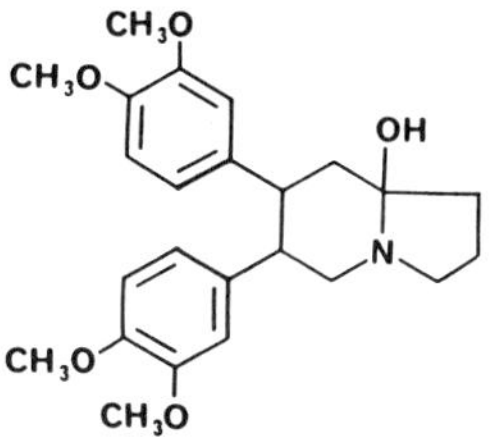

123 13a-Hydroxysepticine

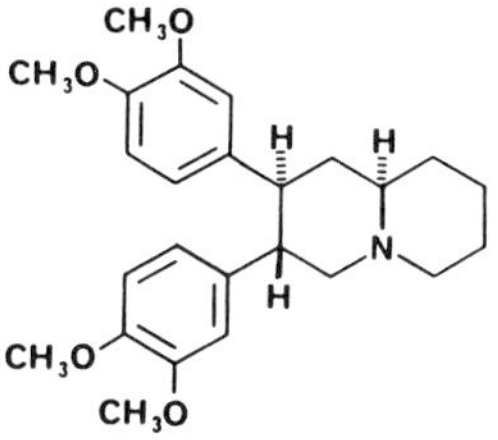

124 Kayawongine

126 and **127**, respectively. Hydrogenolysis and *O*-acylation, followed by *N*-acylation either with 3,4-(dimethoxyphenyl)acetyl chloride or 4-(methoxyphenyl)acetyl chloride, selective ester hydrolysis, and oxidation, led to the β-amino ketone intermediates **128** and **129**, respectively. Intramolecular aldol condensation and photochemical cyclization resulted in a series of isomeric pentacyclic lactams differing from each other in the substitution pattern of rings A and B. Reduction of the lactam carbonyl at an appropriate point in the reaction sequence afforded (±)-tylophorine (**116**) and (±)-cryptopleurine (**117**) from intermediates **130** and **131**, respectively.

j. Benzo[*c*]phenanthridine Alkaloids

Benzo[*c*]phenanthridine alkaloids are widespread in nature, and their occurrence and physical properties have recently been summarized (Krane *et al.*, 1984; Ninomiya and Naito, 1983; Shamma, 1972b; Shamma and Moniot, 1978b; Simanek, 1985). Two

125

126 n=1

127 n=2

128 n=1

129 n=2

130 n=1, R=OCH_3

131 n=2, R=H

representatives, nitidine (**132**) and fagaronine (**133**), are highly active against the P388 lymphocytic leukemia system, although their activity in other test systems is somewhat less.

A new minor alkaloid of *Corydalis ambigua* Cham & Schltd., ambinine (**134**), has been isolated (Ciu *et al.*, 1984). 6-Hydroxymethyldihydronitidine (**135**) has recently been obtained from the stem bark of *Fagaropsis angolensis* (Engl.) Dale (Khalid and Waterman, 1985). Fusion of fagaronine (**133**) afforded *N*-demethylfagaronine (**136**) as the major product, along with the *N*-demethylated compounds **137** and **138**. According to KB and P388 cell culture assays, compounds **136–138** were less cytotoxic than fagaronine (**133**), demonstrating the importance of the quaternary amine moiety for the activity of benzophenanthridines (Arisawa *et al.*, 1984b).

Synthetic efforts aimed at the development of active antitumor benzo[*c*]phenanthridines continue. Since work in this area up to 1985 has been thoroughly reviewed (Shamma, 1972b; Shamma and Moniot, 1978b; Suffness and Cordell, 1985; Ninomiya and Naito, 1983; Simanek, 1985), only a few recent contributions will be mentioned here. Cushman and Chen (1986) have reported the asymmetric synthesis of (+)-corynoline (**139**). Condensation of the chiral ferrocene-type Schiff base (**140**) with an appropriately substituted homophthalic anhydride (**141**) gave **142** with high optical

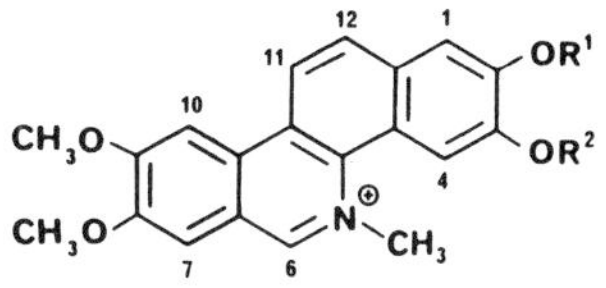

132 Nitidine $R^1+R^2=CH_2$

133 Fagaronine $R^1=H$, $R^2=CH_3$

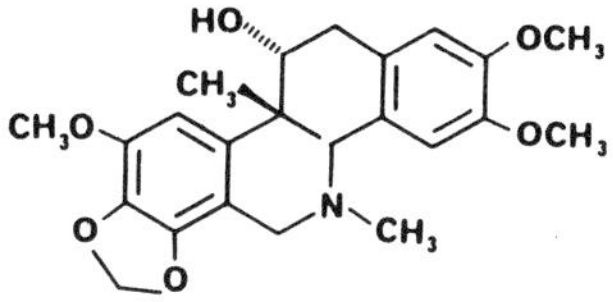

134 Ambinine

135 6-Hydroxymethyldihydronitidine

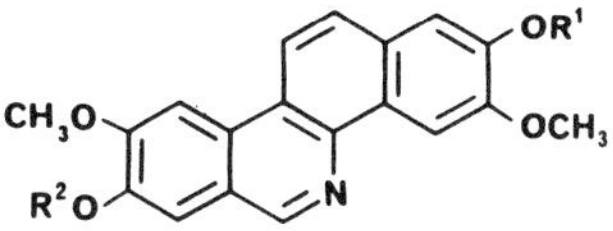

136 $R^1=H$, $R^2=CH_3$

137 $R^1=R^2=CH_3$

138 $R^1=R^2=H$

purity. Removal of the chiral auxiliary group from **142** afforded an intermediate that could readily be transformed into (+)-corynoline (**139**). Utilizing the other enantiomer of **140**, the synthesis of (−)-corynoline has also been achieved.

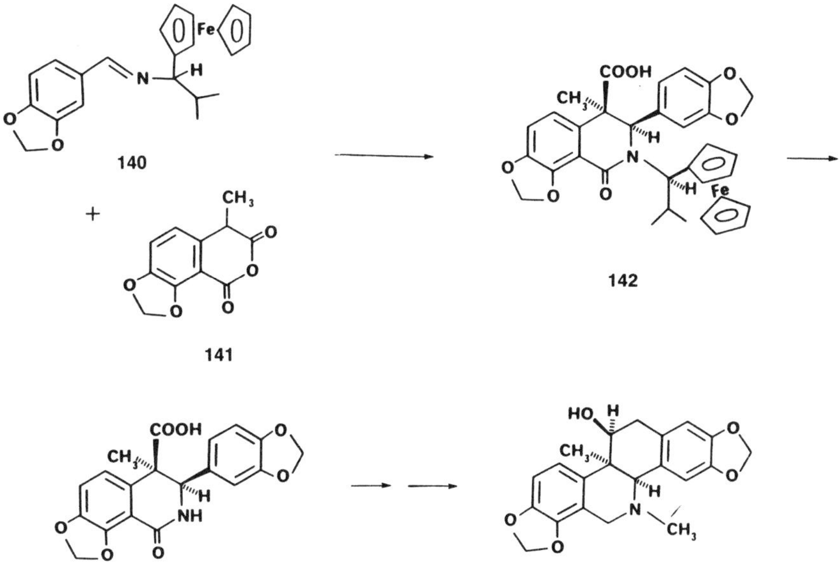

140

141

142

143

139 Corynoline

(±)-Homochelidonine (**144**) was stereoselectively synthesized from berberine (**145**) by Hanaoka *et al.* (1985a). Hofmann elimination of berberine methosulfate (**146**) followed by oxidation with *m*-chloroperbenzoic acid afforded the betaine **147**. Sodium borohydride reduction of **147** and subsequent acid-catalysed cationic cyclization gave benzo[*c*]phenanthridine (**148**) with B/C *cis* anellation. Introduction of the hydroxyl groups at C-11 and C-12 was carried out with performic acid, and finally, the 12-hydroxy group was removed from **149** with triethylsilane in the presence of boron trifluoride etherate to afford **144**.

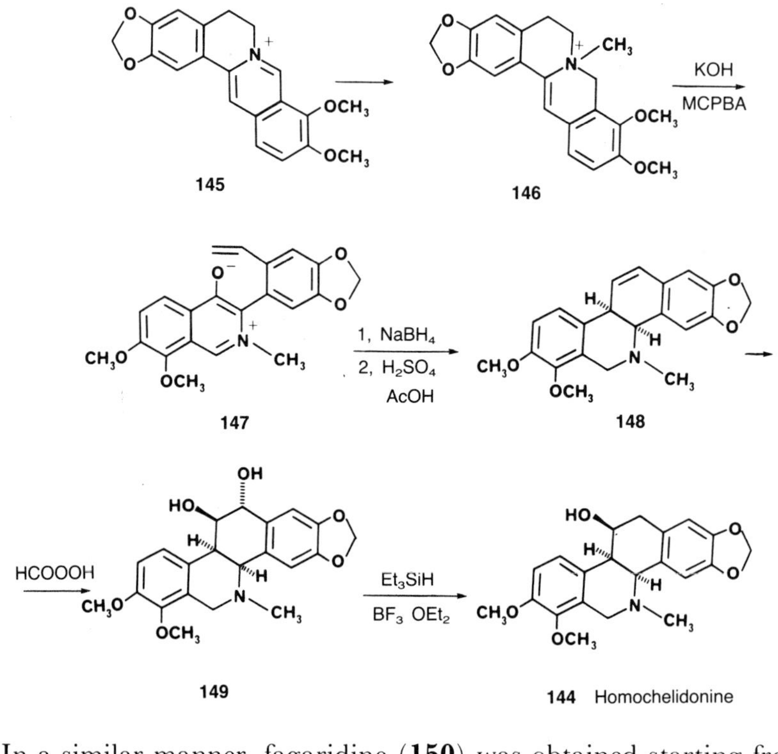

In a similar manner, fagaridine (**150**) was obtained starting from berberrubine (Hanaoka *et al.*, 1985b). (±)-Corydalic acid methyl ester (**151**), a benzo[*c*]phenanthridine derivative lacking the closed C ring, has also been prepared from the protoberberine, corysamine (Hanaoka *et al.*, 1984). The fagaronine analog indenoisoquinoline **152**, as well as its positional isomer **153**, have been prepared and tested against P388 lymphocytic leukemia (Cushman and Mohan, 1985). It has been established that compounds **152** and **153** exhibit

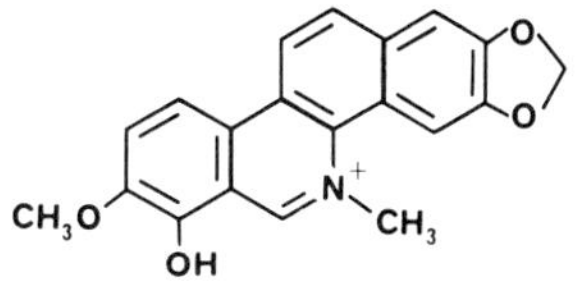

150 Fararidine

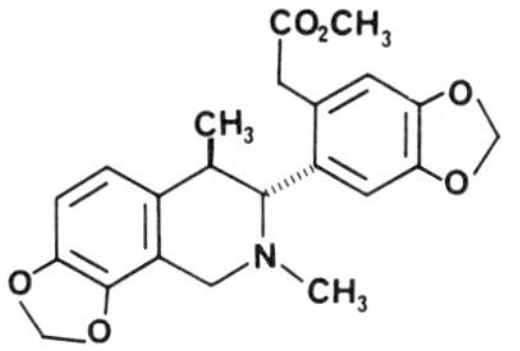

151

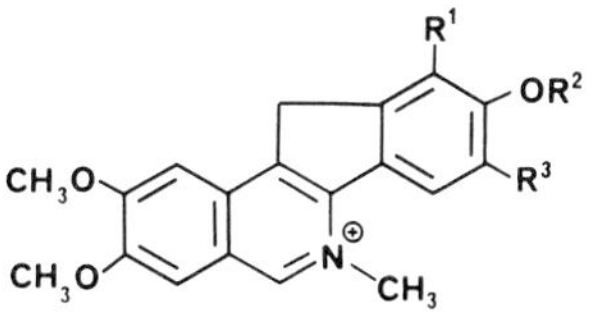

152 $R^1=OCH_3$, $R^2=R^3=H$

153 $R^1=R^2=H$, $R_3=OCH_3$

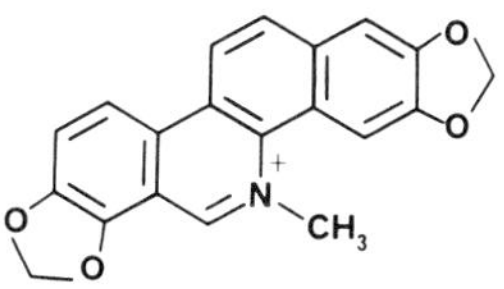

154 Sanguinarine

significant antitumor activity, irrespective of the substitution pattern.

The benzo[*c*]phenanthridine alkaloid sanguinarine (**154**) has been studied for its ability to form a complex with DNA and its inhibitory effect on DNA hydrolysis (Faddejeva *et al.*, 1984). It has been established that sanguinarine (**154**) binds preferentially to the GC pairs in the DNA template (Nandi and Maiti, 1985). The hepatotoxic potential of sanguinarine (**154**) has also been studied (Dalvi, 1985).

k. Amaryllidaceae Alkaloids

Of the skeletal variations of the *Amaryllidaceae* alkaloids, only the lycorine (**155**) type, all of which contain a phenanthridine moiety, display antitumor activity. The *Amaryllidaceae* alkaloids have been reviewed on several occasions (Fuganti, 1975; Grundon, 1984, 1985b), and some of their structure–activity relationships have been discussed (Weng *et al.*, 1982).

A palmitinic acid ester derivative of lycorine (**155**), named palmilycorine (**156**), as well as the acylglucosyloxy alkaloid lycorisidine (**157**), have been isolated from *Crinum asiaticum* L. and their biological effects evaluated (Ghosal *et al.*, 1985a). Compound **157** markedly potentiated the viability of ascites tumor cells, whereas palmilycorine (**156**) had an inhibitory effect. Kalbretorine (**158**) and kalbreclasine (**159**) have recently been isolated from *Haemanthus kalbreyeri* Baker and their pharmacological properties investigated

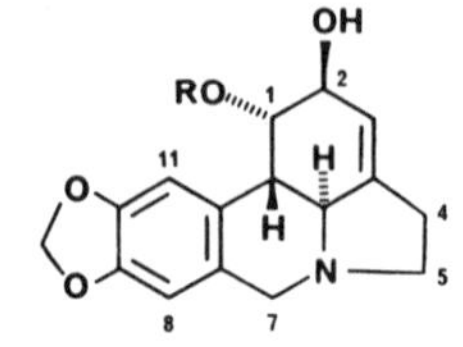

155 Lycorine R=H

156 Palmilycorine R=plamitoyl

157 Lycoriside
R= C′-*O*-palmitoyl-β-*D*-glucopyranosyl

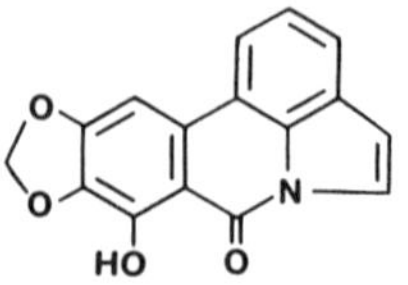

158 Kalbretorine

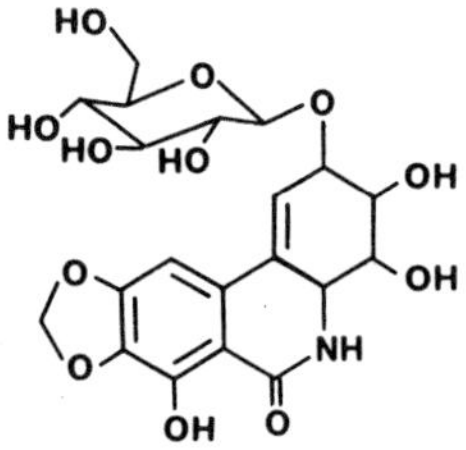

159 Kalbreclasine

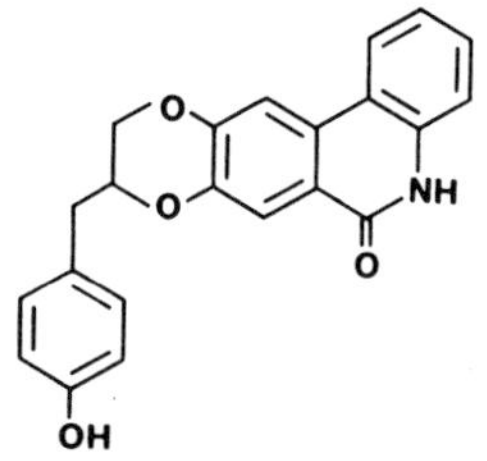

160 Crinasiatine

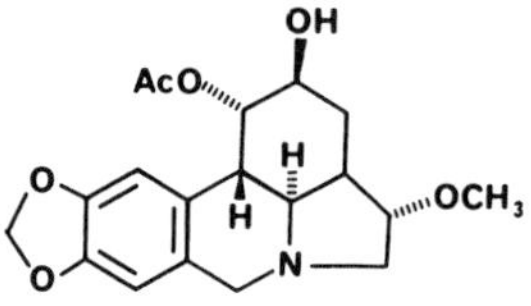

161 Lutessine

(Ghosal *et al.*, 1985b). It has been shown that kalbretorine (**158**) markedly inhibits the growth of S-180 tumor cells, while kalbreclasine (**159**) produces extensive proliferation of splenic lymphocytes in mice.

The first lignanophenanthridine alkaloid, crinasiatine (**160**), has been isolated from the flowering bulbs of *Crinum asiaticum* L. (Ghosal *et al.*, 1985c). It was mentioned (Ghosal *et al.*, 1985c) that **160** exerts both tumor-inhibiting and bacteriostatic activities, but no data have been reported in support of this statement. A new alkaloid lutessine (**161**) and its deacetyl derivative were isolated from the bulbs of *Sternbergia leutea* Ker-Gawl. The structures were elucidated through the interpretation of [^{1}H] and [^{13}C]NMR data (Evidente, 1986).

The structures of the previously known ungeremine (**162**) (lycobetaine) and the recently isolated criasbetaine (**163**), both obtained from fruits of *Crinum asiaticum* L., were determined by chemical and spectroscopic means and confirmed by synthesis starting from lycorine (**155**) and methylpseudolycorine (**164**), respectively (Ghosal *et al.*, 1986). Oxidation of lycorine (**155**) with selenium dioxide resulted in the hydroselenide **165**, which exhibited distinct differences in its UV spectra from that of betaine **162**. Compound **165** could be converted into ungeremine (**162**) by passing its acetone solution over a column of De-Acidite FF at pH 8. Similarly, starting from **164**, the synthesis of criasbetaine (**163**) was achieved (Ghosal *et al.*, 1986).

SeO_2

155 $R^1+R^2=CH_2$ **165** $R^1+R^2=CH_2$ **162** $R^1+R^2=CH_2$

164 $R^1=R^2=CH_3$ **166** $R^1=R^2=CH_3$ **163** $R^1=R^2=CH_3$

I. Bisindole Alkaloids

The clinically and commercially most important antitumor alkaloids are the bisindole alkaloids vincaleukoblastine (**167**) and leurocristine (**168**), isolated from the leaves of the Madagascan periwinkle, *Catharanthus roseus* (L.) G. Don. The successful clinical application of these compounds in the treatment of leukemia and Hodgkin's disease has accorded special importance to this group of alkaloids and their related derivatives. Both vincaleukoblastine (**167**) and leurocristine (**168**) belong to the *Aspidosperma*-cleavamine subgroup of bisindole alkaloids, and this area of alkaloid chemistry developed very rapidly because of the vigorous efforts to isolate new active alkaloids, to study their biosynthesis, to synthesize them from the two component halves, to develop methods for their total synthesis, and to prepare derivatives in order to obtain less toxic and more active substances. The bisindole alkaloids were thoroughly reviewed (Cordell and Saxton, 1981) and therefore only more recent developments will be reported here.

Investigation of the aerial parts of *Catharanthus ovalis* Mgf. resulted in the isolation of 15 bisindole alkaloids (Langlois *et al.*, 1980). Among them, two new *Aspidosperma*-cleavamime type alkaloids have been found, namely, vincovaline (**170**) and vincovalinine (**171**).

Vincovaline (**170**) is an isomer of vincaleukoblastine (**167**) and, according to CD measurements, it possesses the *R* and *S* absolute configurations at stereo centers C-16′ and C-14′, respectively. Vincovalinine (**171**) is the 16′-demethoxycarbonyl derivative of leurosine (**172**) (Langlois *et al.*, 1980).

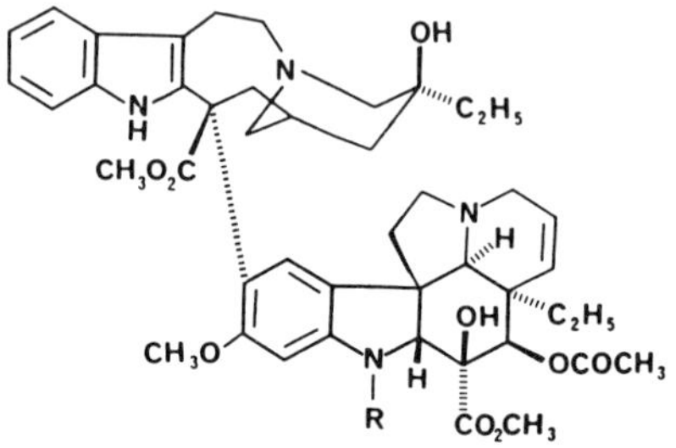

167 Vincaleukoblastine R=CH_3

168 Leurocristine R=CHO

169 Demethylvincaleukoblastine R=H

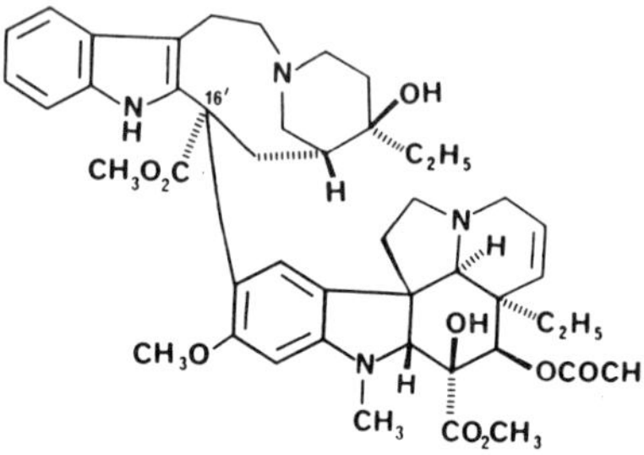

170 Vincovaline

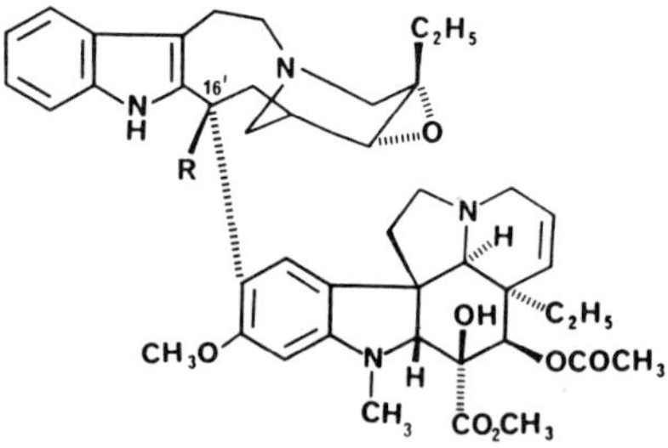

171 Vincovalinine R=H

172 Leurosine R=CO_2CH_3

As part of a continuing study of *Catharanthus* plant species, Cordell and co-workers reported the isolation of 21′-oxoleurosine (**173**) from *Catharanthus roseus* (El-Sayed *et al.*, 1980). The isolate was evaluated for anticancer activity according to established protocols. No activity was observed in the P388 lymphocytic leukemia system in mice, however, **173** exerted cytotoxic activity in the KB test system *in vitro*. An additional minor alkaloid, leurosidine N'_b-oxide (**174**), has also been reported from *C. roseus* (Mukhopadhyay and Cordell, 1981). Leurosidine N'_b-oxide (**174**) was found to be cytotoxic in both the P388 and KB test systems *in vitro*. Catharanthamine (**175**) has also been obtained from *C. roseus* (El-Sayed and Cordell, 1981), and is the first bisindole alkaloid having oxygenation at C-17′ of the velbanamine unit in the form of a novel ether linkage between C-17′ and C-20′. Catharanthamine (**175**) was found to be cytotoxic in the KB test system *in vitro*, and displayed a significant activity in the P388 lymphocytic leukemia test system. Pleurosine, the N′$_b$-oxide of leurosine (**172**), was found to be exceptionally active in the B16 melanoma test system *in vivo* (El-Sayed *et al.*, 1983). Roseadine (**176**), a new isolate of *C. roseus*, has been characterized in detail and its significant activity in the P388 lymphocytic leukemia test system has been demonstrated (El-Sayed *et al.*, 1983).

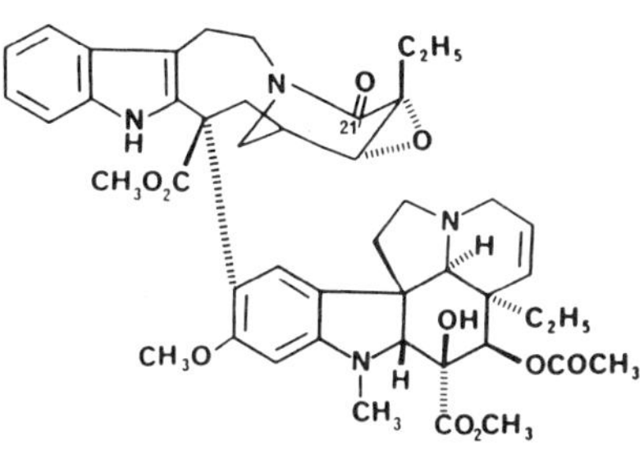

173 21–Oxoleurosine

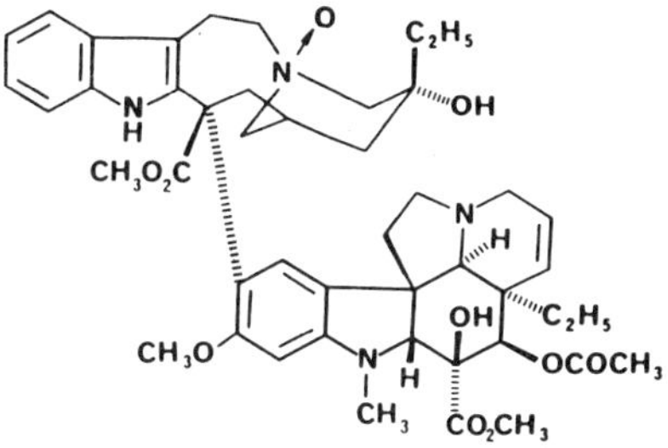

174–Leurosidine *N*-oxide

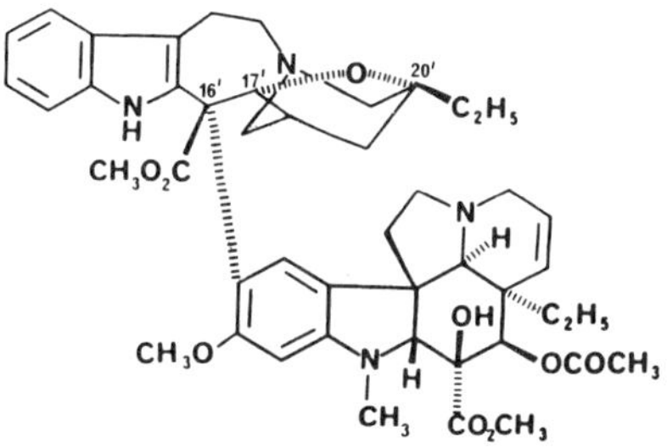

175 Catharanthamine

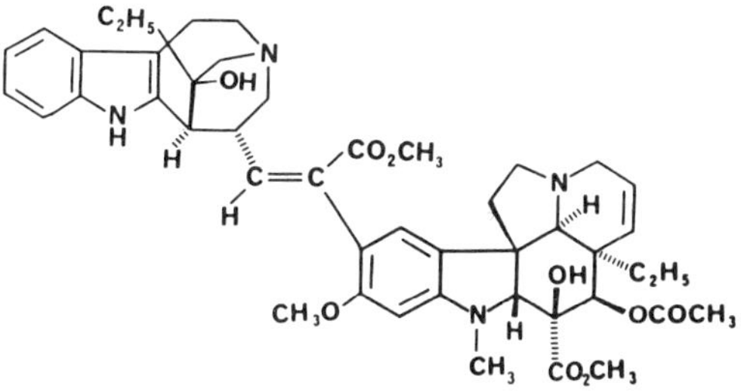

176 Roseadine

Reinvestigation of *C. roseus* has also led to the isolation of 17-deacetoxyvincaleukoblastine (**177**), 17-deacetoxyleurosine (**178**), (De Bruyn *et al.*, 1982), and demethylvincaleukoblastine (**169**) (Simonds *et al.*, 1984). The compounds were identified by [^{1}H]NMR and mass spectrometry, but their biological properties have not been reported.

An efficient method for the conversion of vincaleukoblastine (**167**) to leurocristine (**168**) was developed by a Hungarian group utilizing chromic acid at low temperature in order to obtain larger quantities of **168** from the plant (Jovanovics *et al.*, 1980).

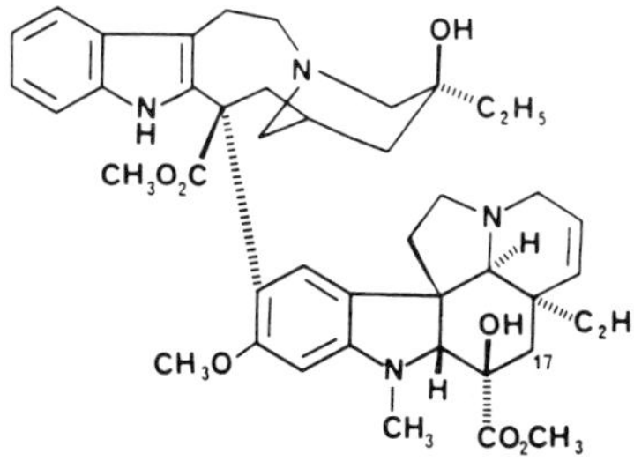

177 17–Deacetoxyvincaleukoblastine

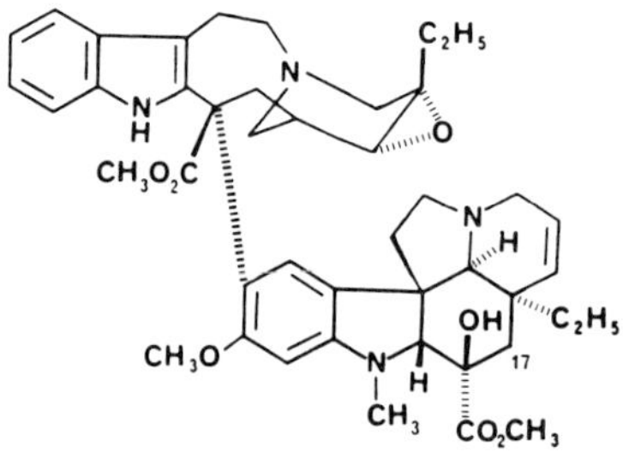

178 17–Deacetoxyleurosine

Since the early 1960s several efforts have been made for the stereoselective coupling of vindoline (**179**) at the C-10 position with the C-16′ position of an appropriate cleavamine derivative. Two methods were generally successful for this coupling: the chloroindolenine approach and the modified Polonovski reaction. Unfortunately, the chloroindolenine approach resulted in a bisindole product with an unnatural absolute configuration at C-16′. The modified

Polonovski reaction, however, led to stereoselective coupling with the correct stereochemistry at C-16′. The groups of Potier (1980a,b), Kutney (1978), and Atta-ur-Rahman (1980) have made substantial contributions to this field and their efforts have been summarized. A comprehensive and critical review (Cordell, 1983) on this subject has appeared.

Most of the recent synthetic efforts have been directed toward the synthesis of analogs or derivatives of vincaleukoblastine (**167**) in order to evaluate highly active, but less neurotoxic anticancer agents. Through the coupling of catharanthine-*N*-oxide (**180**) and vindoline (**179**) induced by a Polonovski fragmentation reaction, anhydrovinblastine (**182**) has been prepared (Langlois *et al.*, 1976). Similarly, the coupling reaction of (±)-20-deethylcatharanthine *N*-oxide (**181**) and vindoline (**179**) led to 20′-deethylanhydrovinblastine (**183**), together with a C-16′ and C-14′ diastereoisomer (Gueritte *et al.*, 1981).

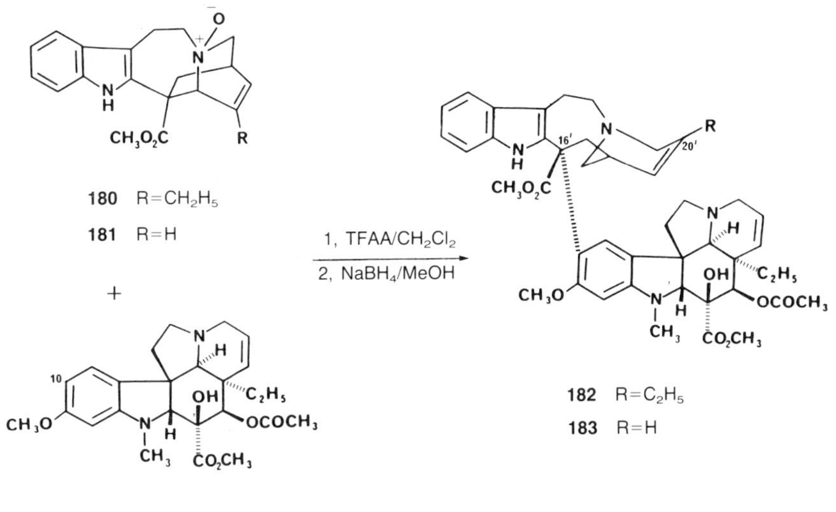

180 $R=CH_2H_5$

181 R=H

+

179

1, $TFAA/CH_2Cl_2$
2, $NaBH_4/MeOH$

182 $R=C_2H_5$

183 R=H

12′-Iodovincaleukoblastine has been prepared from vincaleukoblastine (**167**) by treatment with $Fe(ClO_4)_2$ and tetrapropylammonium-periodate in the presence of RuO_2. 12′-Iodovincaleukoblastine sulfate showed antineoplastic activity against B16 melanoma, P-1334 PJ leukemia, and CA-755 adenocarcinoma (Pearce, 1984). Deacetylvincaleukoblastine acid azide (**184**) was reacted with a series of amino acid ester derivatives to yield vinblastin-23-oyl amino acid derivatives

(**185**). Compounds of type **185** were tested against P388 and L1210 leukemias. The optimum antitumor activity of this series of vincaleukoblastine derivatives was found in the case of isoleucine and tryptophane ethyl esters (Bhushana *et al.*, 1985a,b).

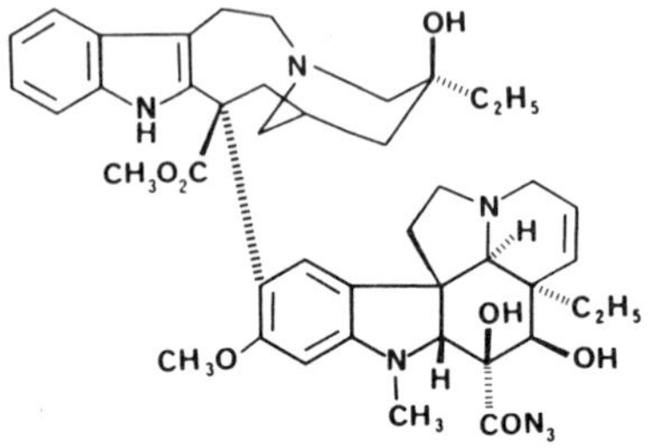

184

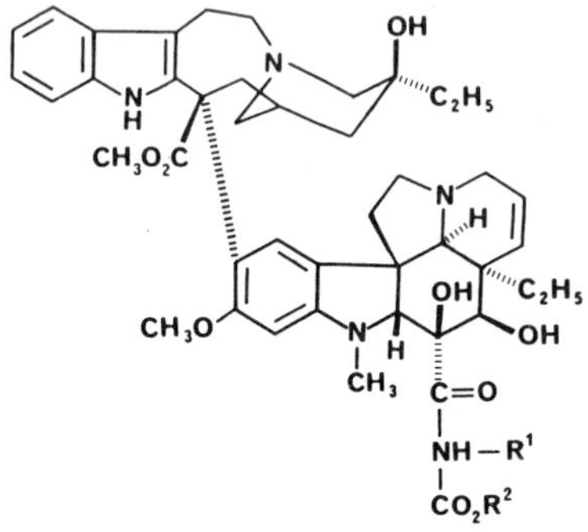

185

Treatment of leurocristine (**168**) with $SOCl_2$ or $POCl_3$ in the presence of DMF gave 15′,20′-anhydroleurocristine (**186**), which on oxidation with O_2 or peroxide resulted in *N*-demethyl-*N*-formylleurosine (**187**). Compound **187** exerted a strong metaphase-blocking effect in HeLa cells *in vitro*, and higher activity against P388 leukemia than that of leurocristine (**168**) itself (Szantay *et al.*, 1982). The antitumor activity of semisynthetic 20′-deoxy-epileurocristine (**188**) has been studied against HxRh 12 rhabdomyosarcoma grown as a xenograft. The efficacy of **188** was lower than that of leurocristine (**168**), but far superior to that of vincaleukoblastine (**167**) in this model (Mullin *et al.*, 1985).

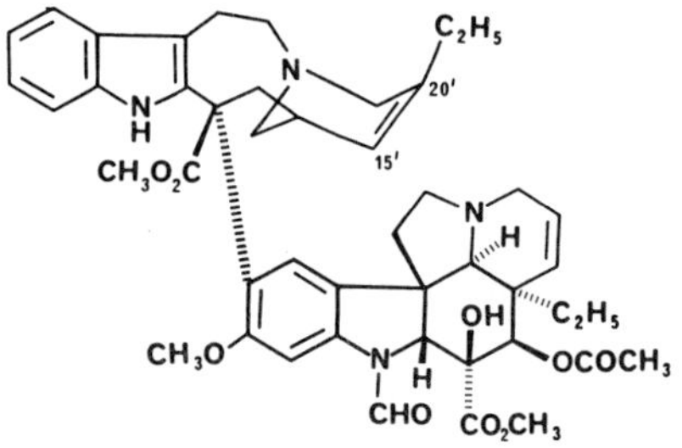

186

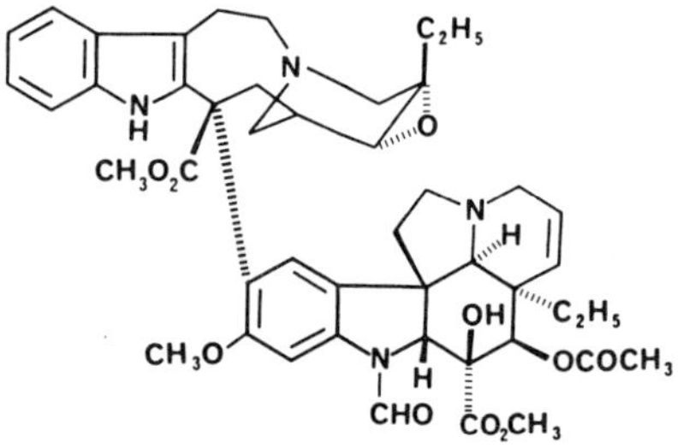

187

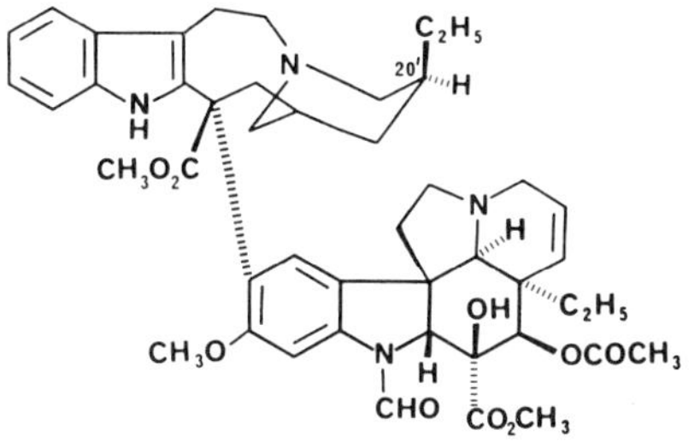

188

A new semisynthetic *Catharanthus* alkaloid analog, navelbine (**189**), has recently been developed and has proved to be as active as leurocristine (**168**) on implanted murine tumors (Maral *et al.*, 1984). Vinepidine (**190**), a new derivative of leurocristine (**168**), together with vincaleukoblastine (**167**), leurocristine (**168**), and vindesine (**191**) were examined for their ability to inhibit microtubule assembly *in vitro* and cell proliferation in culture (Jordan *et al.*, 1985). Results on the pharmacology of vindesine (**191**) (Ueda *et al.*, 1983; Yamamoto *et al.*, 1983) as well as of vinzolidine (**192**) (Kreis *et al.*, 1986), a recently developed derivative, have been reported. The cytogenic effect of vincaleukoblastine (**167**) and leurocristine (**168**) has been studied. It has been established that both alkaloids are powerful spindle poisons, inhibiting cells in metaphase (Thomas-Jacob and Reddi, 1984). A critical review on the pharmacology, toxicology, and pharmacokinetics of vincaleukoblastine (**167**), leurocristine (**168**), and vindesine (**191**) has appeared (Brade, 1981). The effect of *N*-formylleurosine on DNA synthesis has been reported (Pokorny *et al.*, 1983).

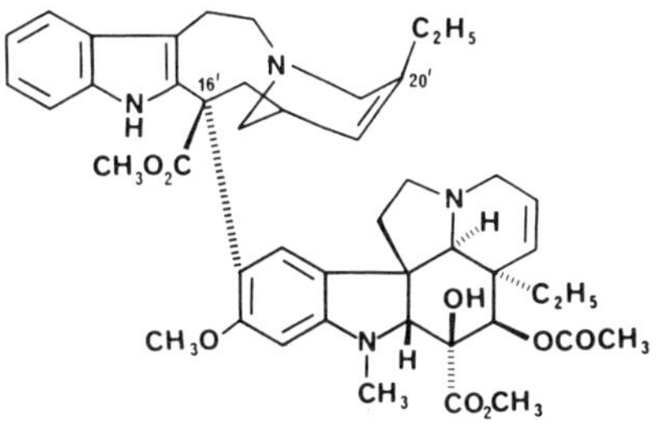

189 Navelbine

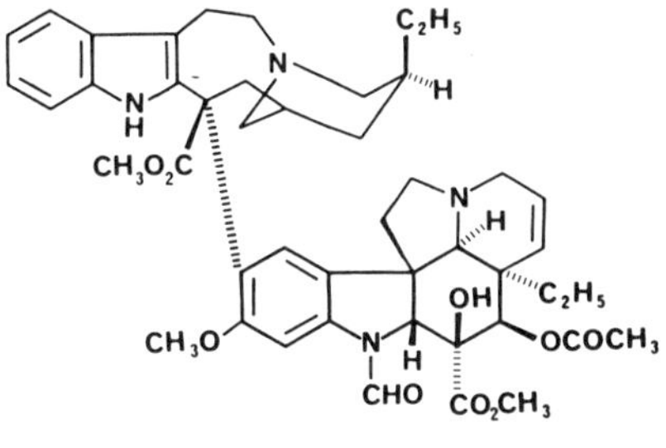

190 Vinepidine

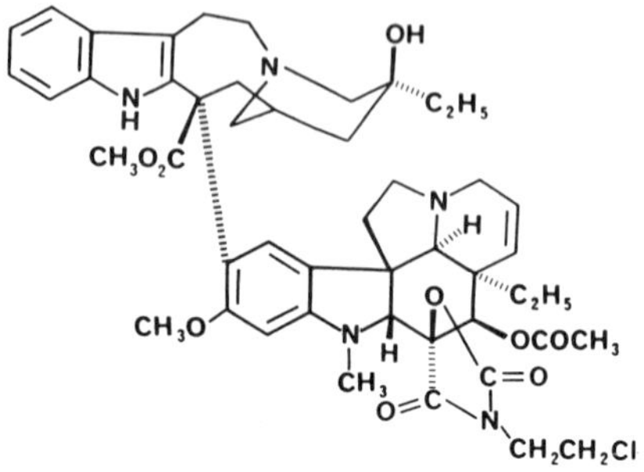

192 Vinzolidine

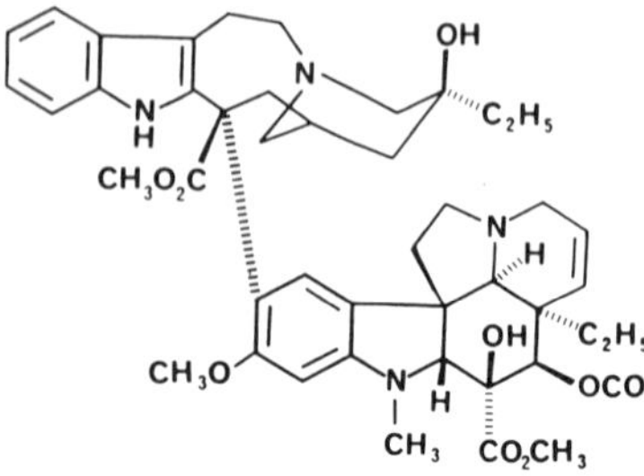

191 Vindesine

m. Maytansinoids

Recent developments in the maytansinoid alkaloid field, including their physical properties, structure elucidation, synthesis, and biological activities, have recently been extensively reviewed (Reider and Roland, 1984; Smith and Powell, 1984).

II. NON-ALKALOIDS

A. LIGNANS

Podophyllotoxin (**193**) and several related aryltetralin lignans, deoxypodophyllotoxin (**194**), 3′-demethylpodophyllotoxin (**195**), α-peltatin (**196**), and β-peltatin (**197**), isolated from the plant families Berberidaceae and Cupressaceae, are cytostatic spindle poisons. Natural podophyllotoxins, such as compounds **193–197**, were unacceptable as antitumor drugs because of their side-effects at higher doses. However, two semisynthetic podophyllotoxin derivatives, etoposide (**198**) and teniposide (**199**), have proved to be antitumor agents with minimal toxic side-effects and are presently undergoing Phase III clinical trials.

A few new members of the podophyllotoxin-type lignans have been isolated and characterized. 1,2,3,4-Tetradehydrodeoxypodophyllotoxin (**200**) was obtained from *Hernandia ovigera* L. (Yamaguchi *et al.*, 1982), while *Haplophyllum davuricum* L. yielded daurinol (**201**) (Batsuren *et al.*, 1981). Justicinol (**202**) has been isolated from *Justicia flava* Vahl. (Olaniyi and Powell, 1980), and podophyllotoxone (**203**) has been obtained both from *Podophyllum hexandrum* and *P. peltatum*

193 Podophyllotoxin
$R^1 = OH$, $R^2 = CH_3$

194 Deoxypodophyllotoxin
$R^1 = H$, $R^2 = CH_3$

195 3′-Demethylpodophyllotoxin
$R^1 = OH$; $R^2 = H$

196 α-Peltatin R=H

197 β-Peltatin $R = CH_3$

198 $R = CH_3$

199 R= (2-thienyl, dihydro)

(Dewick and Jackson, 1981). *Cleistanthus patulus* Muell., Arg. afforded a new lignan glycoside, cleistanthoside A, which was identified as diphyllin-4-*O*-[β-D-glucopyranosyl-(1.2)]-β-3,4-di-*O*-methyl-D-xylopyranoside (Sastry and Rao, 1983).

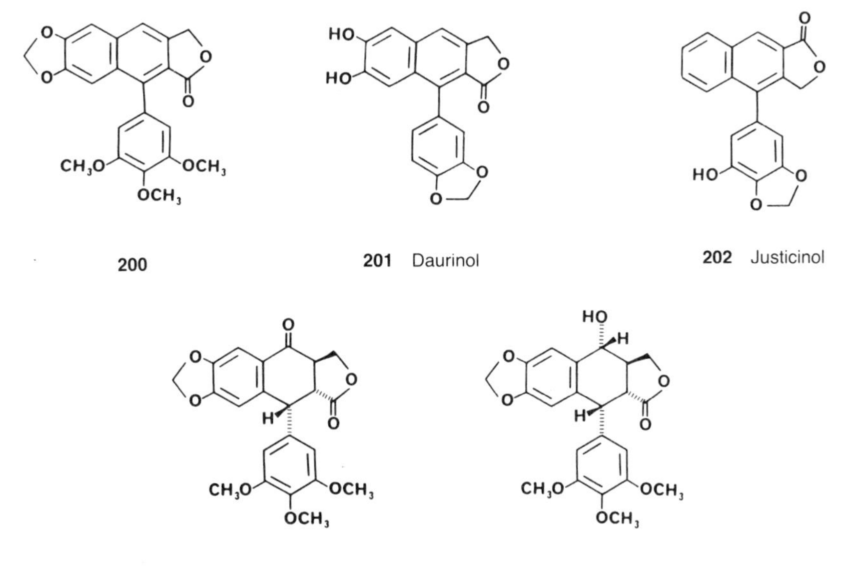

200 **201** Daurinol **202** Justicinol

203 Podophyllotoxone **204** Picropodophyllin

The CD and ORD spectra of a number of 1-aryltetralins have been measured (Hulbert *et al.*, 1981) and [^{1}H]NMR and [^{13}C]NMR investigations of podophyllotoxin (**193**) have been reported (Fonesca *et al.*, 1980; Rithner *et al.*, 1983).

Early efforts on the synthesis of podophyllotoxin (**193**) were based on the formation of picropodophyllin (**204**), which was subsequently converted to its THP-enolate and kinetically reprotonated to podophyllotoxin (**193**) (Gensler and Gatsonis, 1966; Kende *et al.*, 1977; Murphy and Wattanasin, 1980, 1982). The first successful approach to directly synthesize the 1,2-*cis*-2,3-*trans*-substituted aryltetralone moiety was reported (Rajapaksa and Rodrigo, 1981; Rodrigo, 1980). The dimethyl acetal of 6-bromopiperonal was lithiated and reacted with 3,4,5-trimethoxybenzaldehyde to yield **205**. The Diels–Alder reaction of intermediate **207**, generated from **205** by acid treatment with dimethyl acetylenedicarboxylate, resulted in adduct **208**. Hydrogenation of the latter led to the exclusive formation of endo diester **209**. Base-catalysed epimerization of **209**, followed by lithium triethylborohydride reduction, resulted in intermediate **210**. Reductive ring-opening of **210** gave **211** with a 1,2-*cis*-2,3-*trans*-substitution pattern. Removal of the benzylic hydroxyl group at C-4, followed by lactonization, completed the total synthesis of (±)-deoxypodophyllotoxin (**194**) (Rodrigo, 1980). From key-intermediate **211**, the synthesis of (±)-neopodophyllotoxin (**212**) was also achieved (Rajapaksa and Rodrigo, 1981), which had previously been converted to podophyllotoxin (**193**) in two steps (Renz *et al.*, 1965).

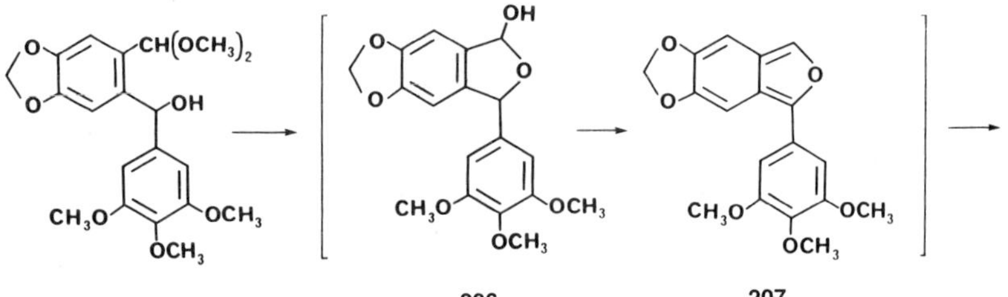

205 206 207

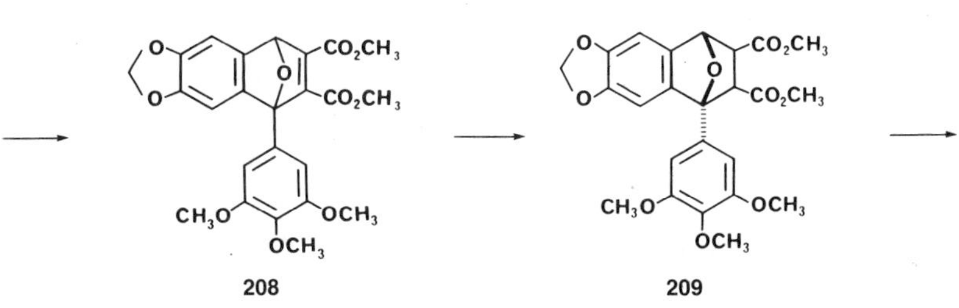

208 209

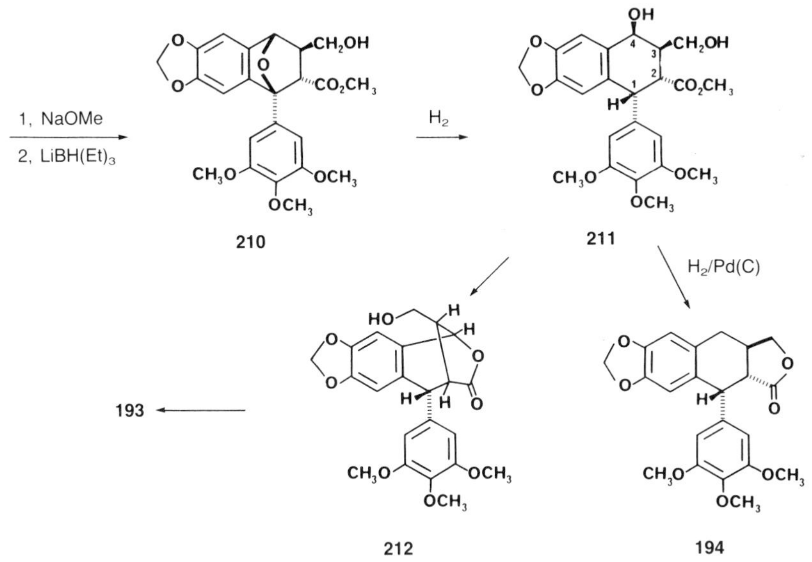

1, NaOMe
2, $LiBH(Et)_3$
210
H_2
211
$H_2/Pd(C)$
193
212
194

A very similar approach was published (Glinski and Durst, 1983) for the synthesis of (±)-epiisopodophyllotoxin (**213**). The Diels–Alder reaction of hydroxyquinodimethane intermediate **215**, generated photochemically from 6-(3′,4′,5′-trimethoxybenzyl)piperonal (**214**) with dimethyl fumarate, resulted in the carbon skeleton of isoepipodophyllotoxin with the correct stereochemistry of the substituents connected to ring B. Adduct **216** was converted in five steps to (±)-epiisopodophyllotoxin (**213**).

An alternative deoxypodophyllotoxin (**194**) synthesis utilizes the Peterson reaction to generate the *o*-quinodimethane intermediate **220**, which, following a Diels–Alder reaction with maleic anhydride, afforded adduct **221**. This compound could be regio- and stereoselectively converted to deoxypodophyllotoxin (**194**) in satisfactory overall yield (Takano *et al.*, 1985).

Several other applications of the Diels–Alder reaction to furnish ring B in aryltetralin lignans have been reported (Das *et al.*, 1983; Mann and Piper, 1982). Additionally, it is worth mentioning a combined oxidation–cycloaddition process (Tsumura Juntendo Co., 1983) which gave, in one step, the podophyllotoxin diastereomer **224** starting from cinnamyl cinnamate (**223**).

α-Hydroxyalkylation of β-(3,4-methylenedioxybenzyl)-γ-butyrolactone (**225**) with 3,4,5-trimethoxybenzaldehyde in the presence of lithium hexamethyldisilylamide gave (±)-podorhizol (**226**) and (±)-epipodorhizol (**227**) as a 1:1 diastereomeric mixture. Cyclization of

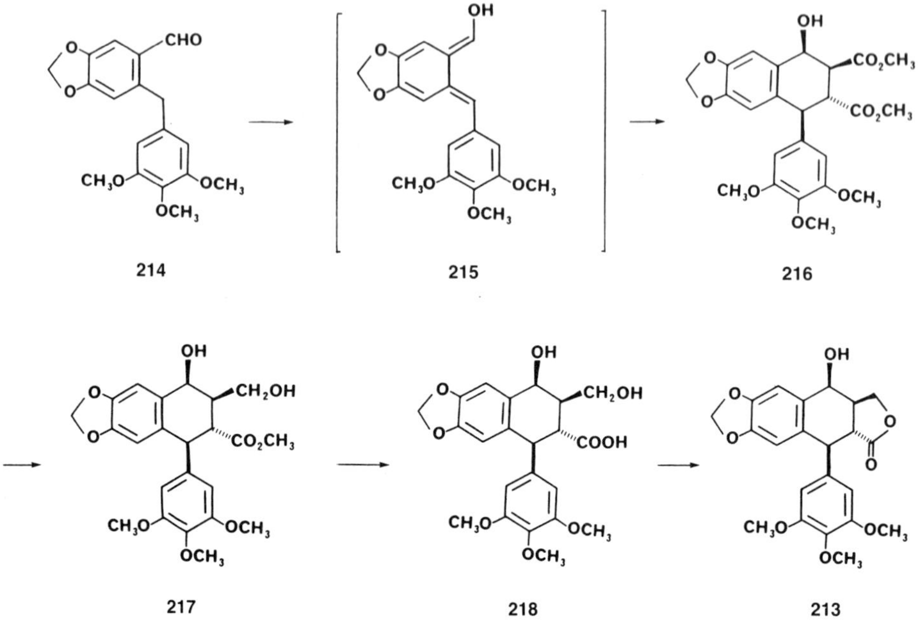

CHO

CH_3O OCH_3 OCH_3

214

OH

CH_3O OCH_3 OCH_3

215

OH CO_2CH_3 CO_2CH_3

CH_3O OCH_3 OCH_3

216

OH CH_2OH CO_2CH_3

CH_3O OCH_3 OCH_3

217

OH CH_2OH COOH

CH_3O OCH_3 OCH_3

218

OH

CH_3O OCH_3 OCH_3

213

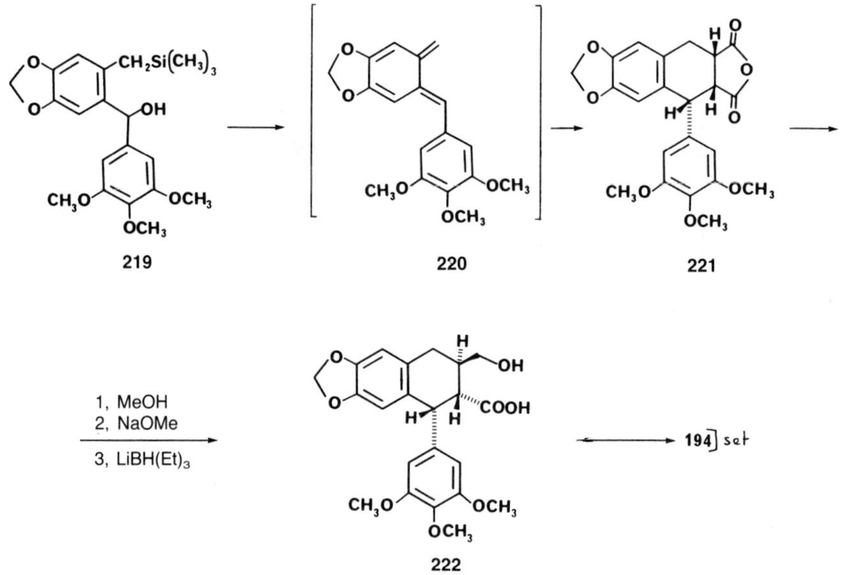

$CH_2Si(CH_3)_3$ OH

CH_3O OCH_3 OCH_3

219

CH_3O OCH_3 OCH_3

220

H H H

CH_3O OCH_3 OCH_3

221

1, MeOH
2, NaOMe
3, $LiBH(Et)_3$

H OH COOH H H

CH_3O OCH_3 OCH_3

222

194] set

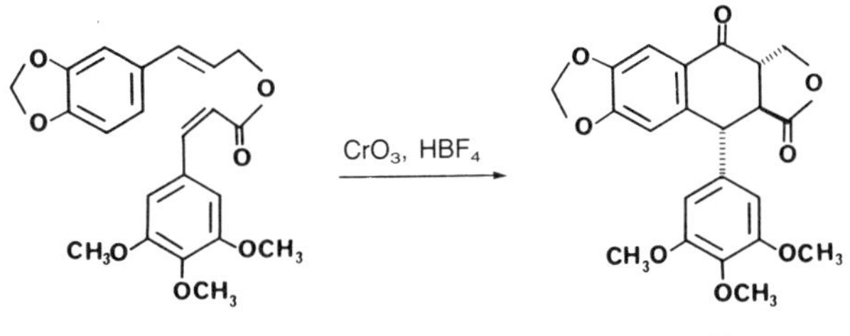

CrO_3, HBF_4

223 **224**

both **226** and **227** using TFA, however, afforded only (±)-isodeoxypodophyllotoxin (**228**) (Robin *et al.*, 1982). (±)-Iso-β-peltatin (**229**) and its methyl ether **230** have similarly been synthesized starting from β-(2-alkoxy-3,4-methylenedioxybenzyl)-γ-butyrolactone (Brown *et al.*, 1982).

The transformation of desoxypodophyllotoxin (**194**) into β-peltatin methyl ether (**231**) and one of its diastereomers has been reported

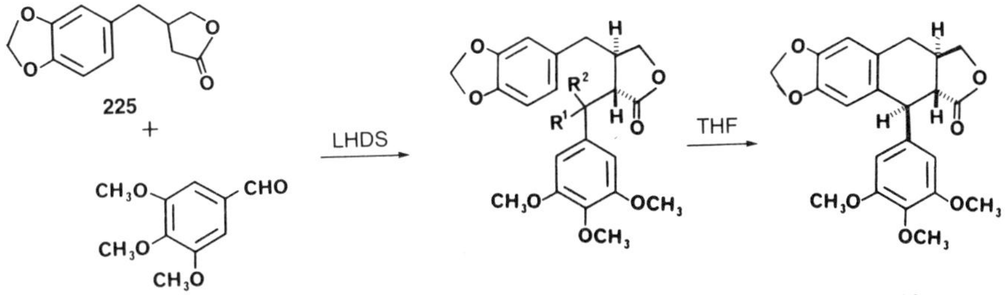

225 + LHDS THF

226 Podorhizol R^1=OH, R^2=H

227 Epipodorhizol R^1=H, R^2=OH

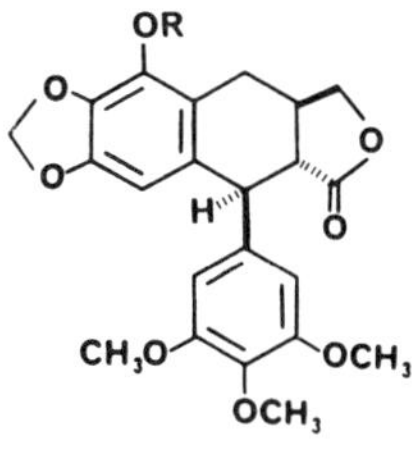

228

229 R=H

230 R=CH_3

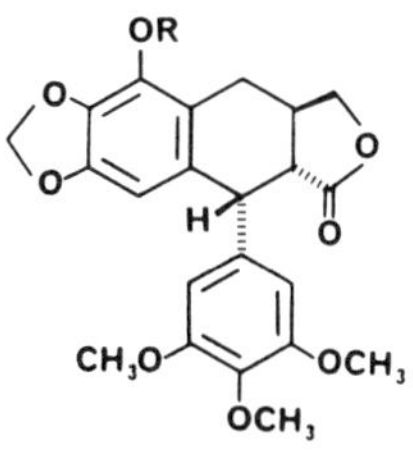

197 R=H

231 R=CH_3

(Yamaguchi *et al.*, 1984) via selective bromination of demethylenedesoxypodophyllotoxin followed by subsequent reconstruction of the methylendioxy substituent and bromine–methoxyl exchange.

A novel approach has been developed by Murphy and Wattanasin (1980, 1982) in which ring B of the aryltetralin moiety is closed by aromatic electrophilic substitution. Fission of cyclopropane derivative **232** generates a carbonium cation that could attack the activated aromatic system, resulting in aryltetralin **233**. Hydrolysis and subsequent hydroxymethylation and oxidative decarboxylation gave (±)-picropodophyllotoxone (**235**).

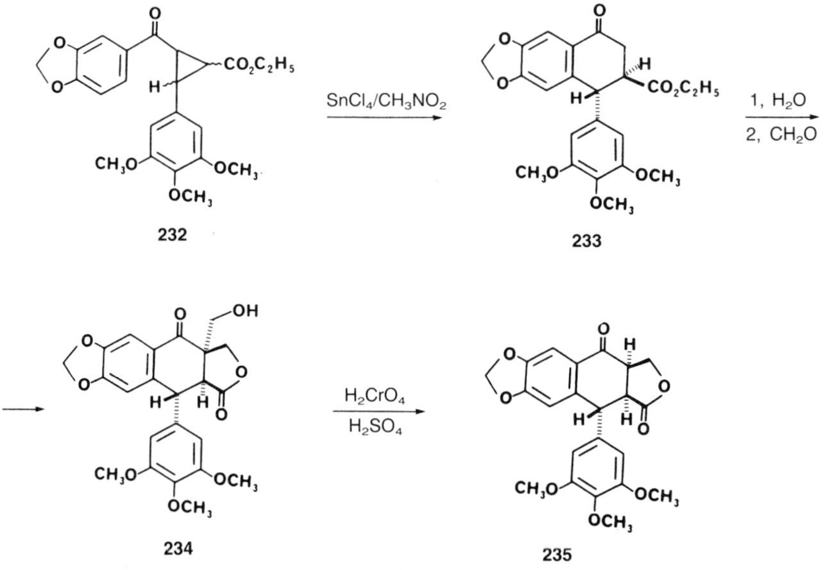

$SnCl_4/CH_3NO_2$; 1, H_2O 2, CH_2O ; H_2CrO_4 H_2SO_4

232 233 234 235

Lignans and lignan analogs with an aromatic ring B have also been prepared (Meyers and Avila, 1981; Plaumann *et al.*, 1980; Wang and Ripka, 1983). A comprehensive review on lignans and neolignans and covering the literature up to 1983 has appeared (Whiting, 1985).

Several reviews have been published on the pharmacological activity, mechanism of action, pharmacokinetics, and clinical pharmacology of podophyllotoxin (**193**) (Loike, 1984; Morufushi, 1984), etoposide (**198**) (Adachi *et al.*, 1985; Creaven, 1982, 1983; D'Incalci *et al.*, 1982; Issell *et al.*, 1982; Loike, 1984; Long and Brattain, 1983; Morita *et al.*, 1986; Newlands, 1983; Phillips and Lauper, 1983; Sinkule, 1984; Winograd *et al.*, 1984), and teniposide (**199**)

(Broggini *et al.*, 1982; Creaven, 1982; Issell *et al.*, 1982; Long and Brattain, 1983; Morita *et al.*, 1986; Winograd *et al.*, 1984). Podophyllotoxin (**193**) is an inhibitor of microtubule assembly both *in vitro* and *in vivo*. Interestingly, in contrast with podophyllotoxin, etoposide (**198**) and teniposide (**199**) do not inhibit the microtubule assembly, probably due to the presence of the bulky glucoside moiety. The cytotoxicity of **198** and **199** could be related to their inhibitory activity on topoisomerase, resulting in an accumulation of DNA breaks (Long *et al.*, 1985). Etoposide (**198**) is the most active agent yet tested for the treatment of small-cell carcinoma of the bronchus, and is also used in the treatment of lung cancer. Additionally, it has therapeutic value against the AIDS-associated Kaposi's sarcoma. In combination with leurocristine (**168**), etoposide (**198**) shows synergistic antitumor activity (Jackson *et al.*, 1984).

Steganone (**236**), first isolated by Kupchan and co-workers, is an example of a dibenzocyclooctadiene-type lignan exerting antitumor activity (Kupchan *et al.*, 1973). This group of lignans comprises several stereoisomers due to the restricted rotation of the biphenyl unit to the different chiral centers on the cyclooctadiene moiety. A substantial number of gomisins (gomisin A–gomisin Q) have recently been isolated from the fruits of *Schisandra chinensis* Baill., which are used therapeutically as antitussives (Ikeya *et al.*, 1978, 1979a,b,c,d,e,f, 1980a,b,c, 1982a,b; Gottlieb *et al.*, 1982). Kadsurin (**237**) and related compounds were isolated from *Kadsura japonica* Benth. (Ookawa *et al.*, 1981), and araliangine (**238**) was obtained from *Steganotaenia araliacea* Hochst. (Taafrout *et al.*, 1983). One new dibenzocyclooctadiene lignan (**239**), an inseparable mixture of two others (**240**, **241**), all with a unique oxygen-bridge linking C-5 and C-8, have been isolated from *Clerodendron inerme* Gaertn. (Spencer and Flippen-Anderson, 1981). Spirodienones, eupodienone-1, -2, and -3 (**242**, **243**, **244**, respectively), have been found in *Eupomatia laurina* Hook (Bowden *et al.*, 1980). These lignans could be intermediates in the biogenesis of dibenzocyclooctadiene-type compounds.

(−)-Steganone (**236**) has been synthesized by Larson and Raphael (1982). Coupling of 3,4-methylenedioxyphenylzinc chloride (**245**) with aryl iodide **246** resulted in the biphenyl aldehyde **247**, which was converted into the phenanthrone derivative **249** by simple steps. Ring expansion followed by chiral resolution gave optically active **250**. Formation of the lactone ring attached to the dibenzocyclooctadiene moiety gave (+)-isosteganone (**251**), which was thermally epimerized into (−)-steganone (**236**).

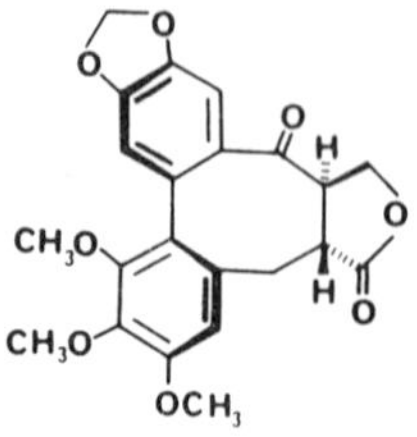

236 Steganone

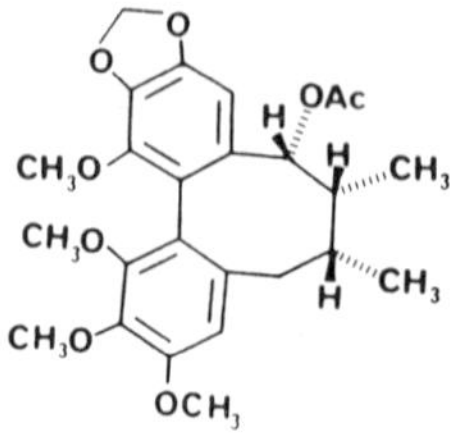

237 Kadsurin

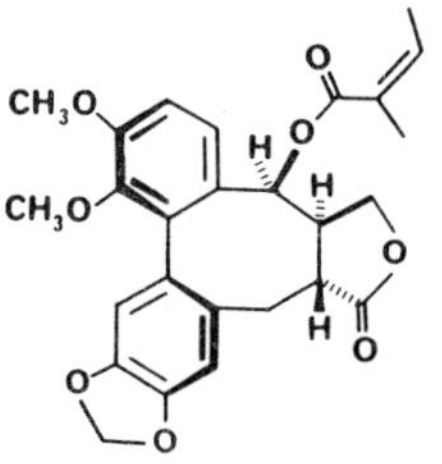

238 Araliangine

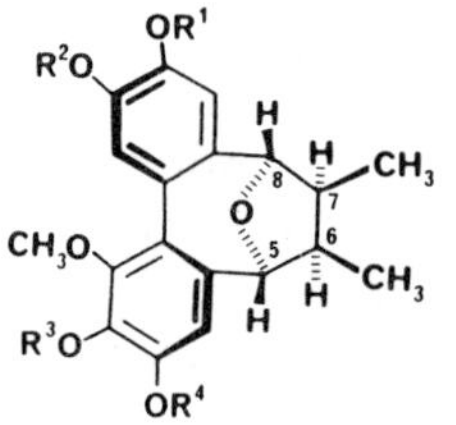

239 $R^1+R^2=R^3+R^4=CH_2$

240 $R^1=R^2=CH_3$, $R^3+R^4=CH_2$

241 $R^1+R^2=CH_2$, $R^3=R^4=CH_3$

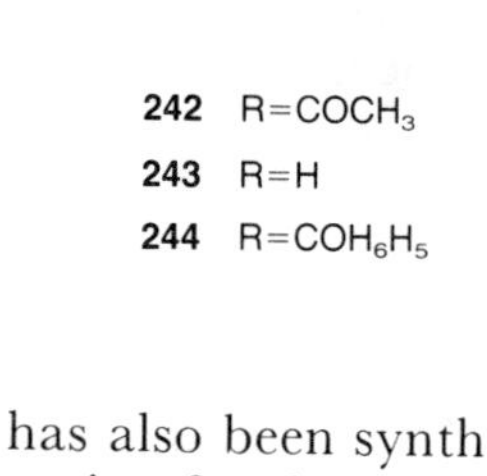

242 $R=COCH_3$

243 $R=H$

244 $R=COH_6H_5$

(±)-Steganone (**236**) has also been synthesized (Dhal *et al.*, 1983) utilizing the Ullmann reaction for the formation of key-intermediate **254**. Biphenyl **254** was cyclized by intramolecular hydroxyalkylation to the isomeric alcohols **255** and **256**. These were subsequently oxidized by Jones' reagent, decarboxylated using barium hydroxide, and reoxidized to afford the isomeric ketoacids **257** and **258**, each

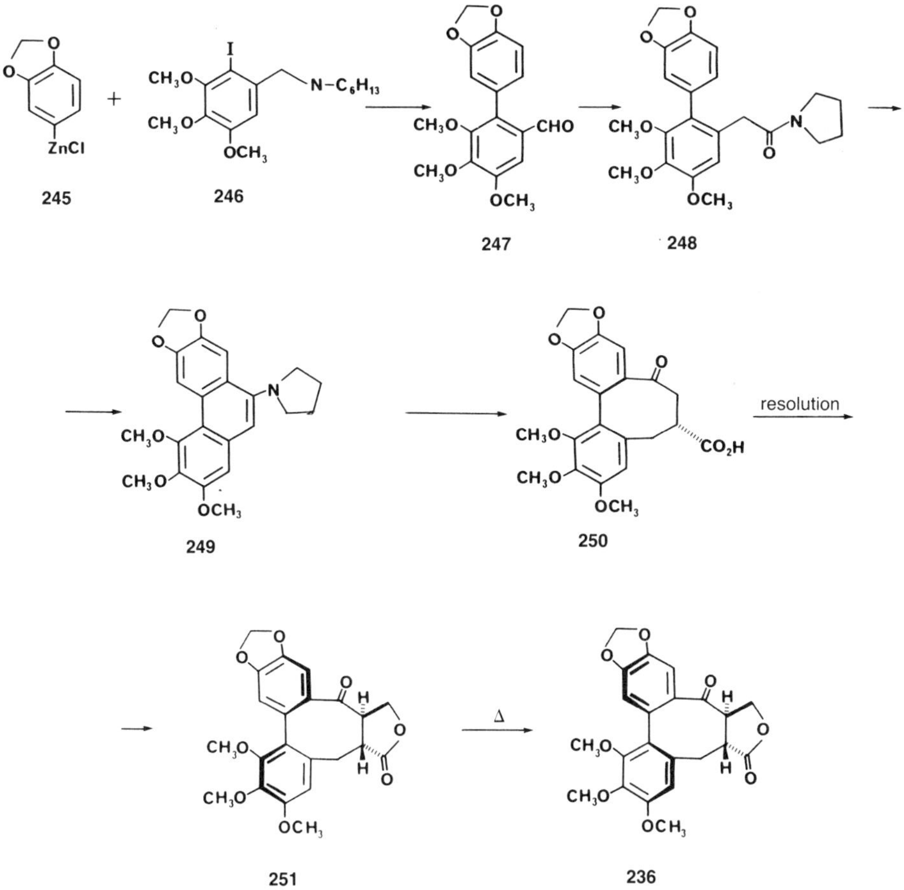

ZnCl
245
246
CHO
247
248
249
250
resolution
CO_2H
Δ
251
236

of which was converted to (±)-steganone (**236**) using Raphael's method as described above.

The stereoselective total synthesis of (±)-steganone (**236**) has been achieved via a thallium trifluoracetate-mediated biaryl coupling reaction (Magnus *et al.*, 1984, 1985). Treatment of phosphonate **259** with sodium hydride and piperonal gave unsaturated ester **260**. The coupling reaction of **260** was performed by thallium trifluoracetate in trifluoroacetic acid followed by a lithium aluminum hydride reduction to afford **261**. Simmons-Smith cyclopropanation of **261** gave **262**. Solvolysis of **262** in glacial acetic acid resulted in ring expansion, and the regio- and stereospecific introduction of the

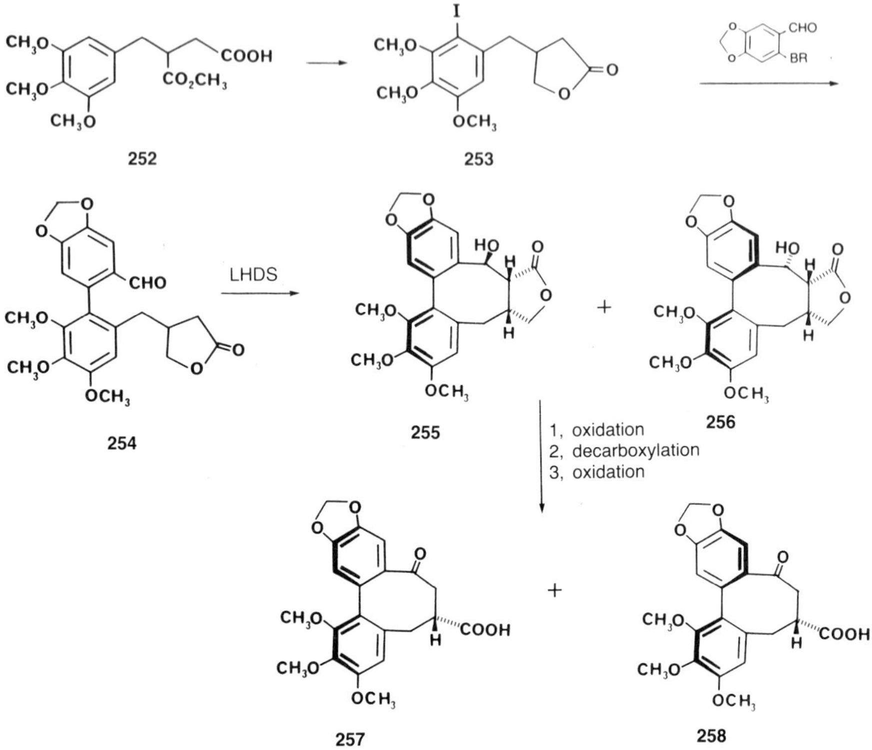

C-8 oxygen substituent gave **263**, which on hydroboration led to diol **264**. Jones oxidation of **264** gave the corresponding keto acid **265**, which was later converted into (±)-steganone (**236**) by standard literature procedures.

Due to the biologically successful modifications of podophyllotoxin (**193**) made through the insertion of a glucopyranosyl moiety into the molecule, attempts have been made to prepare glucoside derivatives of steganol (**259**) similar to etoposide (**198**). Thus 4′,6′-*O*-ethylidene- and thienylidene-β-D-glucopyranosyl derivatives (**260** and **261**, respectively) of steganol (**259**) have been synthesized and evaluated for activity in P388 leukemia *in vivo* and DNA breakage assays (Hicks and Sneden, 1985; Houlbert *et al.*, 1985). None of the compounds proved to be active, and it has been hypothesized that the observed inactivity may be due to the presence of the C-11 methoxy group or possibly the rigid conformation of the biaryl ring system.

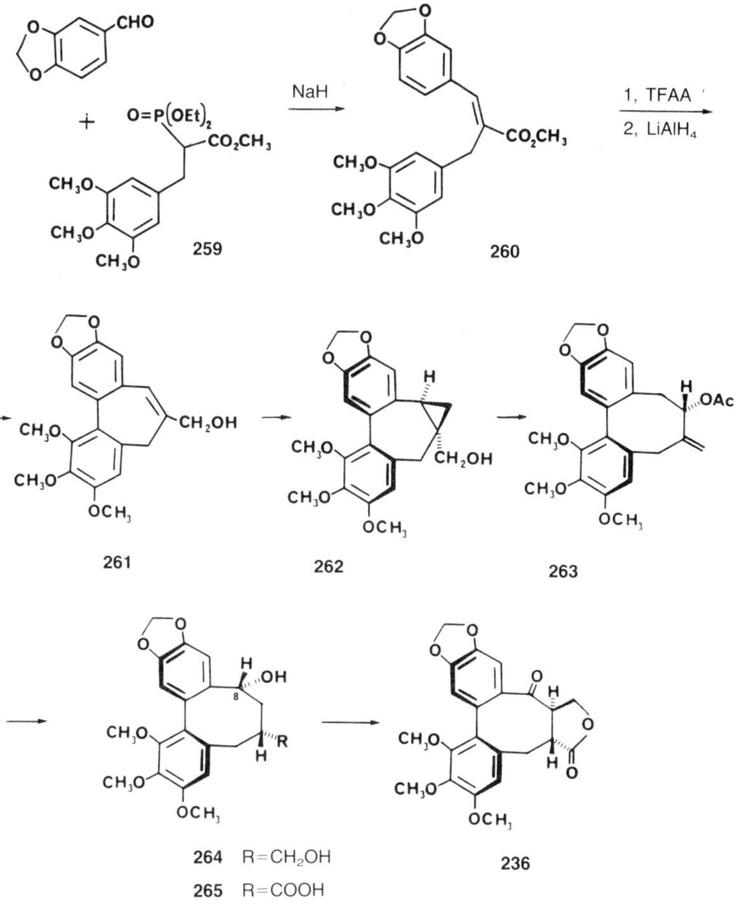

CHO
O=P(OEt)2
CO2CH3
CH3O
259
NaH
CO2CH3
260
1, TFAA
2, LiAlH4
CH2OH
OCH3
261
262
263
OAc
OH
264 R=CH2OH
265 R=COOH
236

B. QUASSINOIDS

Quassinoids, also known as simaroubolides, are a group of degraded triterpene lactone derivatives limited in distribution to the plant family Simaroubaceae. Since the chemistry and biological activity of the quassinoids have been reviewed on several occasions (Polonsky, 1973, 1983, 1985; Cassady and Suffness, 1980), only a limited discussion will be presented here, focussing particularly on those members of this group that exert antitumor activity. There are five basic quassinoid skeleta, but thus far only compounds with the C_{20} skeleton (**266**) have demonstrated antileukemic activity.

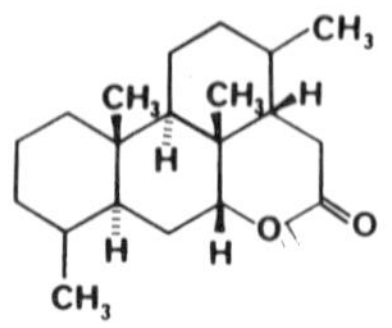

266

A new quassinoid, 6α-acetoxypicrasine B (**267**), has recently been isolated from the stem bark and leaves of *Soulamea fraxinifolia* Brongn. & Gris (Charles *et al.*, 1986). 15-*O*-Benzoyl-brucein D (**268**) has been obtained from *Soulamea amara* Lam. and its structure established from its spectral data and by single-crystal X-ray crystallographic analysis (Bhatnagar *et al.*, 1985). 15-Deacetylsergeolide (**269**), isolated from the leaves of *Picrolemma pseudocoffea* Ducke, exerted strong antileukemic activity in the murine P388 lymphocytic leukemia test system *in vivo*, but it displayed only moderate cell growth inhibition against the P388 cell line *in vitro* (Polonsky *et al.*, 1984). Bruceane (**270**) was isolated from the fruit of *Brucea javanica* Merrill and has demonstrated activity against KB cells *in vitro* (Zhang *et al.*, 1984). Reinvestigation of the Ethiopian tree *Brucea antidysenterica* Mill. led to the isolation and characterization of two additional new antileukemic isobruceine-type quassinoids, bruceanol-A (**271**) and bruceanol-B (**272**). Both compounds showed significant activity against the P388 lymphocytic leukemia system *in vivo* (Okano *et al.*, 1985a). Odyendane (**273**) and odyendene (**274**) have been isolated from the trunk bark of *Odyendea gabonensis* (Pierre) Engl., and their structures were elucidated through spectral data and X-ray analysis (Forgacs *et al.*, 1985). Three bitter quassinoids, yadanziolides A, B, and C (**275**, **276** and **277**, respectively), were isolated from the seeds of *Brucea javanica* Merrill, known as "Ya-dan-zi" in Chinese folk medicine (Yoshimura *et al.*, 1984, 1985). From the same plant species, eight new antileukemic quassinoid glycosides, yadaziosides A–H, were also obtained in addition to two new quassinoids, dehydrobrusatol (**278**) and dehydrobruceantinol (**279**) (Sakaki *et al.*, 1984, 1985). Shinjulactone D (**280**) and shinjulactone E (**281**) have been isolated from *Ailanthus altissima* Swingle (Furuno *et al.*, 1984). Seven new quassinoid glucosides, picrasinosides A–G, as well as two new quassinoid hemiacetals, picrasinol-A (**282**) and picrasinol-B (**283**), have been found in the stem bark of *Picrasma ailanthoides* Planchon and their structures established by spectral analysis (Okano *et al.*, 1984, 1985b). New quassinoid glucosides have also been found in *Simarouba glauca* DC. (Bhatnagar *et al.*, 1984) and *Hannoa klaineana* Pierre (Lumonadio and Vanhaelen, 1985).

Extensive studies have been conducted on the synthesis of C_{20}-quassinoids due to their potent antileukemic and cytotoxic properties.

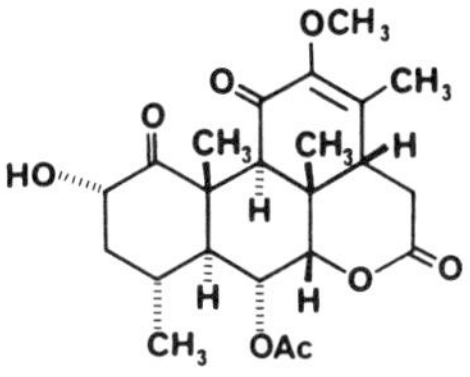

267

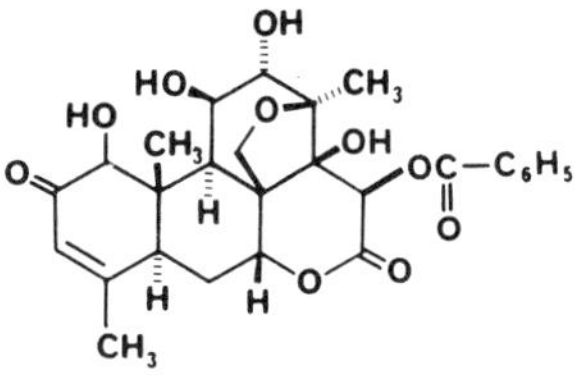

268

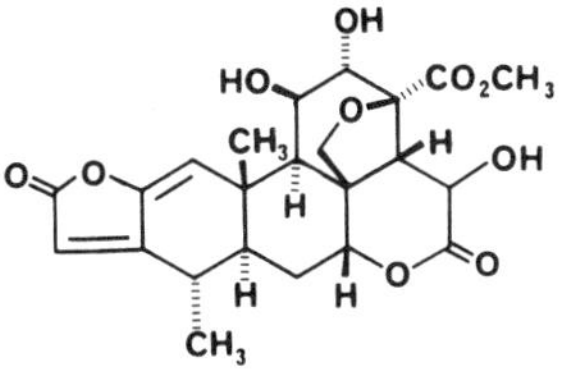

269

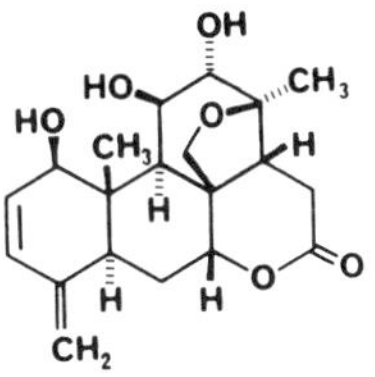

270

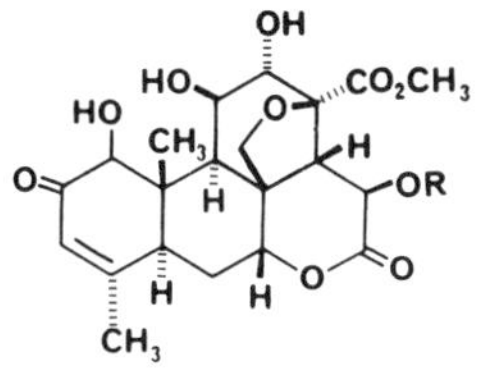

271 $R=COC_6H_5$

272 $R=COC_5H_{11}$

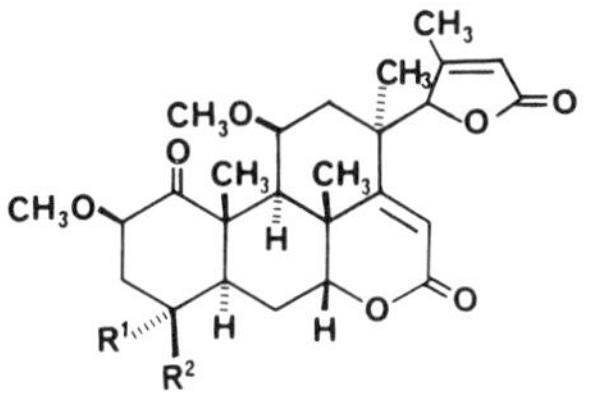

273 $R_1=CH_3$, $R^2=H$

274 $R^1+R^2=CH_2$

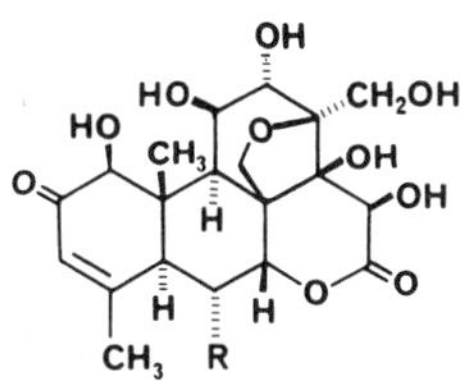

275 R=H

276 R=OH

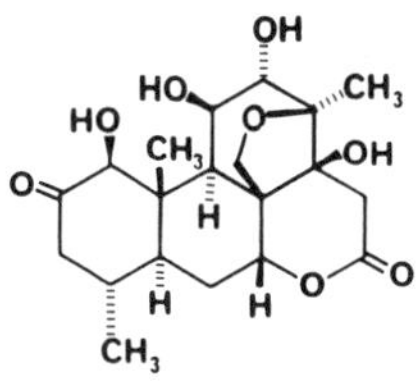

277

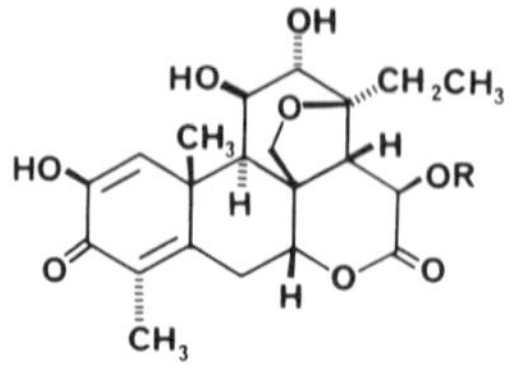

278 R =

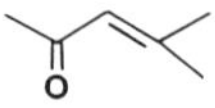

279 R =

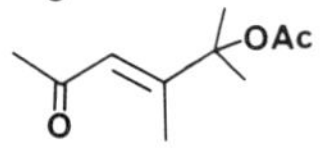

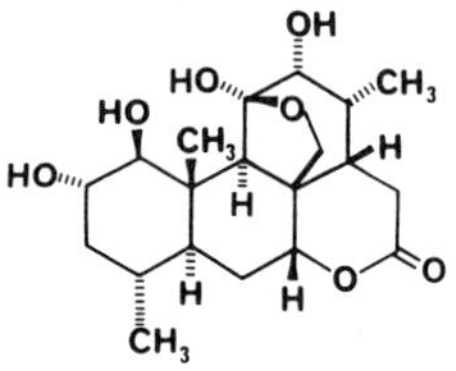

280

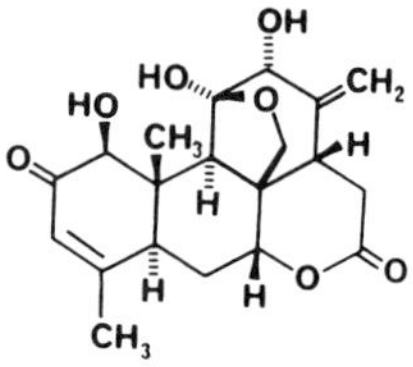

281

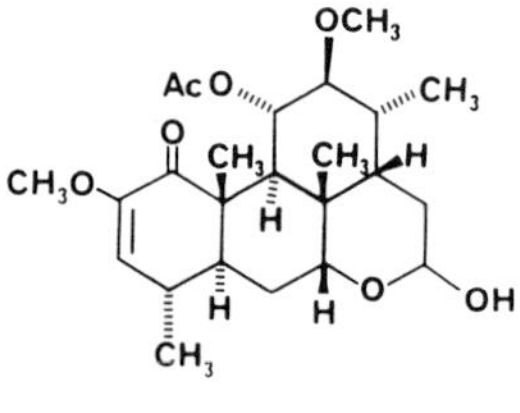

282

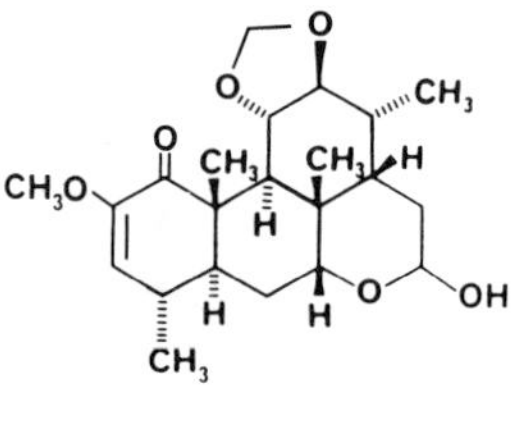

283

A number of approaches toward their synthesis have appeared. However, as a reflection of their stereochemical complexity, total synthesis has so far been limited to the synthesis of (±)-quassin (**284**) and (±)-castelanolide (**285**), as reported by Grieco and coworkers in preliminary (Grieco *et al.*, 1980a, 1982) and detailed studies (Grieco *et al.*, 1984a; Vidari *et al.*, 1984).

Cycloaddition of dienophile **286**, prepared previously (Grieco *et al.*, 1980b) with ethyl (*E*)-4-methyl-3,5-hexadienoate (**287**) in the presence of ethyl aluminum dichloride, gave a single Diels–Alder product, the tricyclic keto ester **288**. Reaction of **288** with sodium borohydride proceeded stereospecifically. Thus, lactonization and demethylation afforded the hydroxy lactone **289**, diisobutyl aluminum hydride reduction of which afforded a corresponding lactol, and treatment with methanolic hydrogen chloride resulted in the protected lactol **290**. Hydroboration followed by Collins oxidation yielded the diketone **291**. Oxygenation of the dianion derived from **291** with MoO_5PH gave bis-(α-hydroxy ketone) **292**, which was transformed into diosphenol **293** with complete inversion of the stereochemistry at C-9. Compound **293** was methylated to afford neoquassin β-*O*-methyl ether (**294**), which on selective hydrolysis of the protected lactol, followed by oxidation with Fetizon's reagent, yielded (±)-quassin (**284**).

Grieco *et al.* (1982, 1984a) have also accomplished the total synthesis of (±)-castelanolide (**285**). Tetrahydropyranylation of the previously prepared key-intermediate **290**, followed by hydroboration of the C-12,13 double bond, gave alcohol **295**. Benzylation of the hindered C-12 hydroxyl group, followed by cleavage of the C-1 tetrahydropyranyl ether and subsequent Collins oxidation, gave ketone **296**. Phosphorylation of the enolate generated from **296**, and simultaneous cleavage of the C–O bonds of the phosphorodiamidate ester and the benzyl ether groups, yielded tetracyclic olefin **297**. Collins oxidation of **297** provided a ketone whose enolate was treated with the MoO_5PH reagent to afford hydroxy ketone **298**. Treatment of **298** with sodium methoxide in methanol and DMSO resulted in oxidation of the C-11 hydroxyl group to a ketone, together with complete inversion of the configuration at C-9, yielding diosphenol **299**. Hydrolysis of **299**, followed by oxidation with Fetizon's reagent, afforded the tetracyclic lactone **300**. Acetylation of the enolate, followed by oxidation of the C-1,C-2 double bond with osmium tetroxide, and removal of the acetyl-protecting group gave crystalline (±)-castelanolide (**285**). This synthesis also provided confirmation of the structure of castelanolide as determined previously (Mitchell *et al.*, 1971).

A stereoselective synthesis of 12β-hydroxypicrasan-3-one (**301**) has been developed (Miyaji *et al.*, 1984). Treatment of the previously prepared tricyclic intermediate **302** (Honda *et al.*, 1981) with triethyl

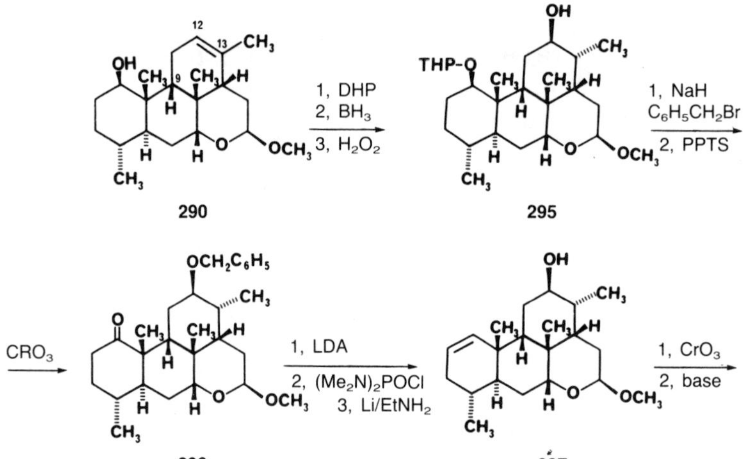

290 → (1, DHP; 2, BH_3; 3, H_2O_2) → 295 → (1, NaH, $C_6H_5CH_2Br$; 2, PPTS; CrO_3) → 296 → (1, LDA; 2, $(Me_2N)_2POCl$; 3, Li/EtNH$_2$) → 297 → (1, CrO_3; 2, base) →

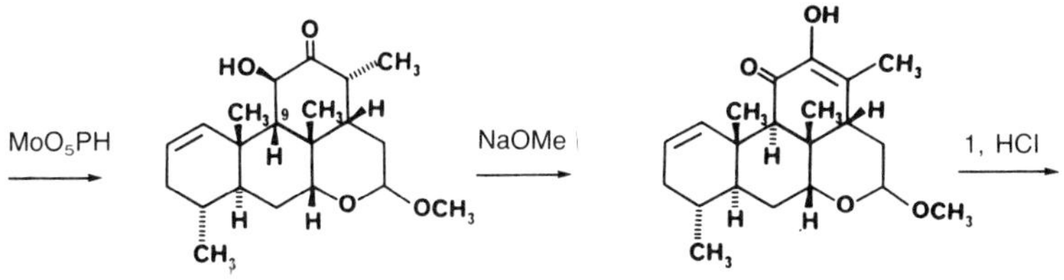

orthoacetate and pentachlorophenol gave **303** via an orthoester Claisen rearrangement. Compound **303** was then stereoselectively converted into **304** through five unexceptional reaction steps ($LiAlH_4$ reduction, acetylation, CrO_3 oxidation, catalytic reduction, and deacetylation). Treatment of **304** with lead tetraacetate and irradiation with a visible lamp resulted in the formation of an oxidative ether linkage, yielding the pyran derivative **305**. Tetracyclic intermediate **305** was transformed by successive reaction steps into 12β-hydroxypicrasan-3-one (**301**), having the correct stereochemistry at all six ring-stereo centers of the picrasane skeleton.

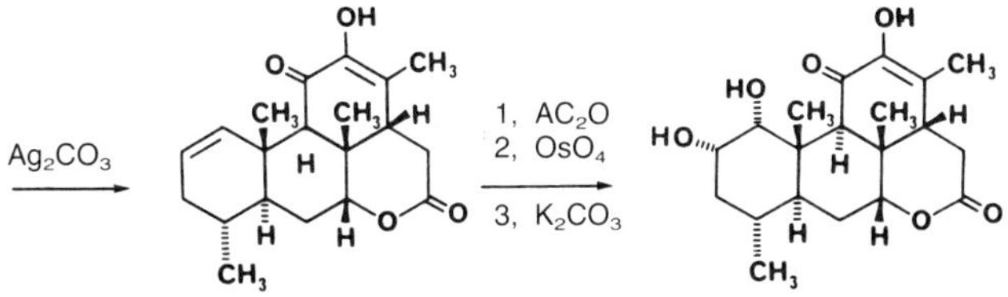

Shishido *et al.* (1984) have synthesized the pentacyclic compound **306** as a model for bruceantin (**307**), an antitumor quassinoid from *B. antidysenterica*. The main feature of the synthesis was a stereoselective intramolecular Diels–Alder reaction of the benzocyclobutane derivatives **308** and **309** into tetracyclic ketones **310** and **311**, respectively. Cycloadducts **310** and **311** were transformed into pentacyclic lactone **306** via a multistep reaction sequence.

Additional studies were conducted (Suryawashi and Fuchs, 1986; Ziegler *et al.*, 1985) in order to establish model experiments for an eventual synthesis of bruceantin (**307**).

A new linear synthetic route to the skeleton of the quassinoids has also been developed (Heathcock *et al.*, 1984). Two-stage annelation of 2-methylcyclohexanone with 1-chloropentanone gave tricyclo dienone **314**, which could be readily oxidized by acetyl chromate to the corresponding dienedienone **315**. Bisketalization,

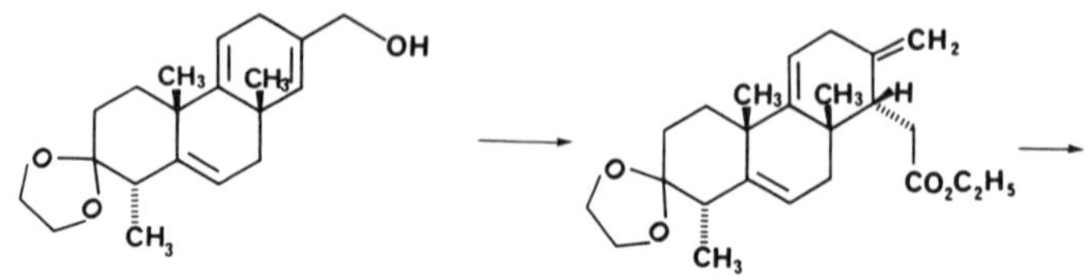

302

303

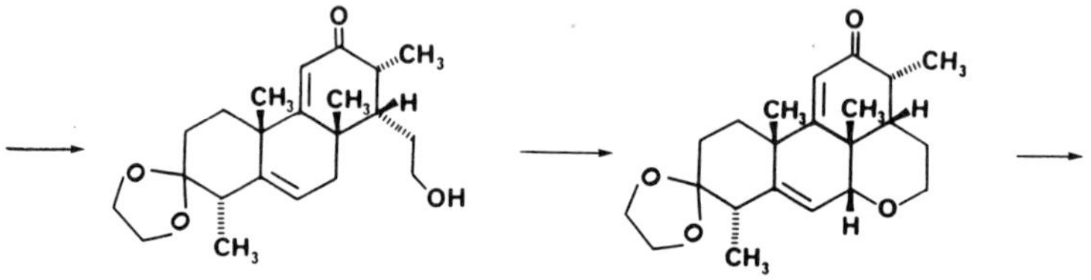

304

305

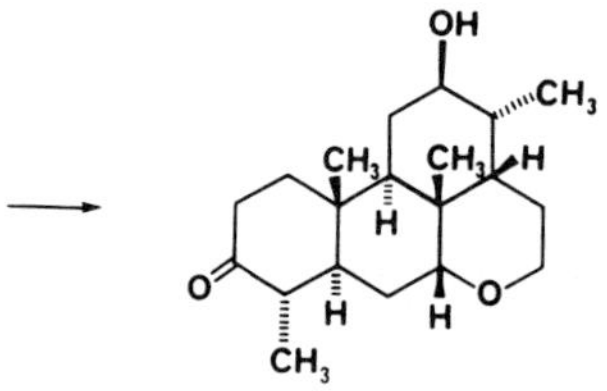

301

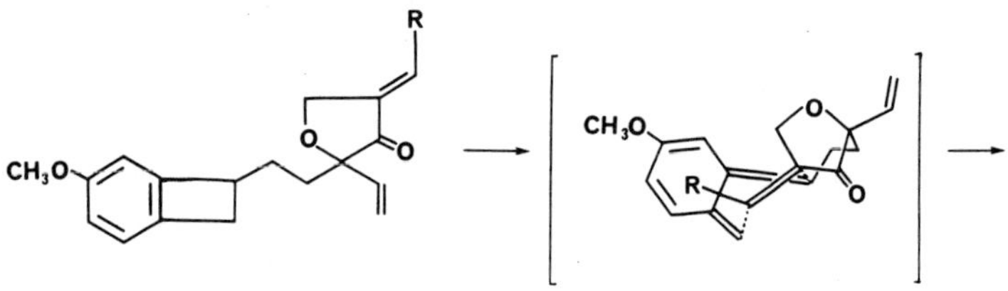

308 R=S−C_6H_5

309 R=OAc

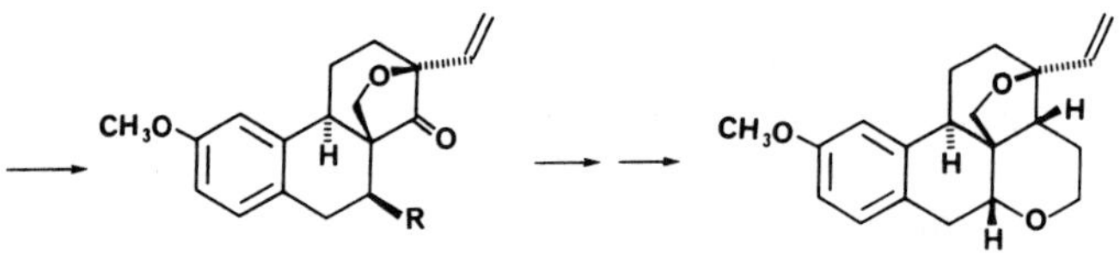

310 R=S−C_6H_5

311 R=OAc

306

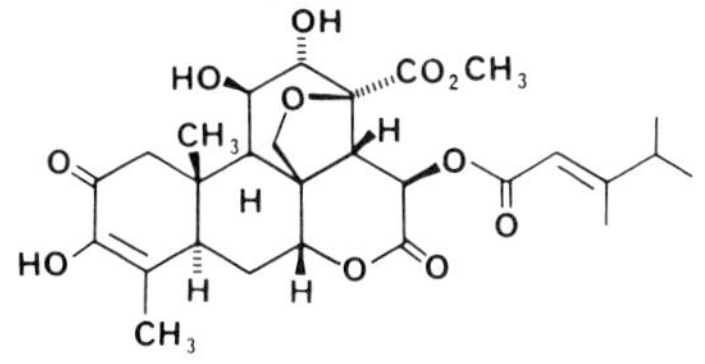

307 Bruceantin

double-bond transposition, and selective hydrolysis provided monoketal **316**, along with recovered **315**. Ring C of **316** was functionalized with the introduction of a methoxycarbonyl group alpha to the carbonyl. The resulting keto ester **317** was transformed into the unsaturated keto ester **318**, which could be reacted with the unsymmetrical ketene acetal **319** under pressure at room temperature. Desilylation of the adduct with aqueous KF afforded keto diester **320**. Epoxidation of **320**, followed by subsequent allylic alcohol formation and pyridinium-chlorochromate-induced solvolytic cyclization, afforded the tetracyclic lactone **322**, which could potentially serve as a key-intermediate for the synthesis of a series of quassinoids.

Novel methods for the construction of the C-8,C-13 epoxymethano bridge of quassimarin and related quassinoids have been developed (Grieco *et al.*, 1984b; Kanai *et al.*, 1984). Synthetic studies on the samaderins have recently been reported (Inanga *et al.*, 1985) in order to achieve the necessary ring contraction of a δ-lactone moiety into a γ-lactone.

With respect to the pharmacological properties of the C_{20}-quassinoids, new information has been obtained on the protein-synthesis-inhibiting activity of bruceantin (**307**). It appears that **307** binds tightly to ribosomes and that this is the major mechanism by which it inhibits protein formation (Willingham *et al.*, 1984). The effects of the quassinoids brusatol and bruceantin on protein synthesis in a number of murine tumors has been studied (Hall *et al.*, 1983). The inhibitory effects of ten natural quassinoids and two semisynthetic analogs were tested against promastigotes of *Leishmania donovani* growing in culture (Gero *et al.*, 1985). Seven of the compounds had 50% inhibitory concentrations between 0.5 and 1.85 μM. The most active compounds were simalikalactone D (**323**) and the semisynthetic derivative **324**; however, both compounds were also toxic to macrophages.

C. MISCELLANEOUS COMPOUNDS

The male antifertility drug gossypol (**325**) (Lei, 1982; Prasad and Diczfalusy, 1982; Segal, 1950) was demonstrated to be a specific inhibitor of DNA synthesis. The antitumor activity of gossypol (**325**) was established first in 1975 by Jolad *et al.* (1975). The effects of gossypol (**325**) on tumor growth and on the survival of mice bearing the mammary adenocarcinoma 755 or the P388 or L1210 leukemias have been investigated (Rao *et al.*, 1985). It has also been

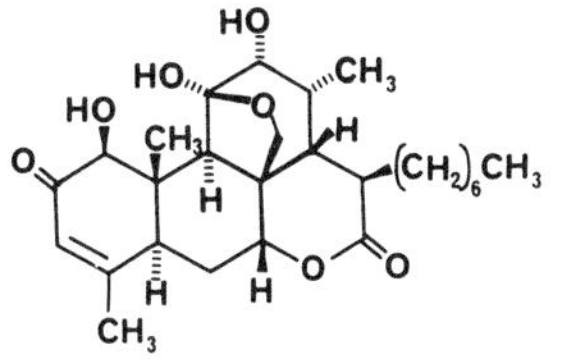

323

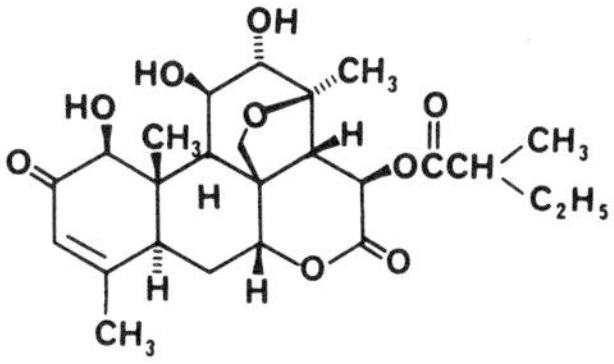

324

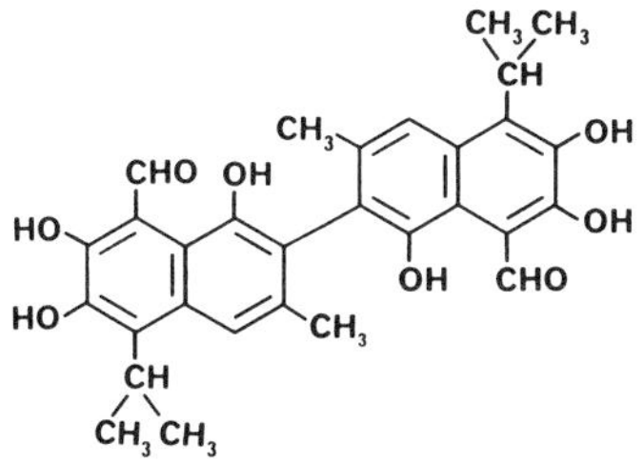

325 Gossypol

demonstrated that gossypol (**325**) inhibits the proliferation of Ehrlich ascites tumor cells (Tso, 1984).

Uvaricin (**326**), a member of a new class of cytotoxic linear acetogenins, has been isolated from *Uvaria accuminata* Oliver (Jolad *et al.*, 1982). Rollinicin (**327**) and isorollinicin (**328**), bearing similar bistetrahydrofuran and α,β-unsaturated γ-lactone moieties, have been obtained from the roots of *Rollinia papilionella* Diels (Dabrah and Sneden, 1984). Both rollinicin (**327**) and isorollinicin (**328**) exhibited cytotoxicity against the P388 lymphocytic leukemia system *in vitro*. Asimicin (**329**), an extremely cytotoxic acetogenin, has recently been isolated from the bark and seeds of the pawpaw tree, *Asimina triloba* Dun. (Rupprecht *et al.*, 1986).

Ostopanic acid (**330**) has recently been isolated from the stems and fruits of *Ostodes paniculata* Blume (Hamburger *et al.*, 1987). It has also been demonstrated that ostopanic acid (**330**) exerted an ED_{50} of 1.5 μg/ml in the P388 lymphocytic leukemia test system *in vitro*.

Two new phloroglucinol derivatives, mallotophenone (**331**) and mallotochromene (**332**), were isolated from the pericarps of *Mallotus japonicus* Muell. Arg. (Arisawa *et al.*, 1985). Both **331** and **332**, together with other known phloroglucinol derivatives isolated from this tree, were found to be cytotoxic against the KB and L5178Y systems in cell culture.

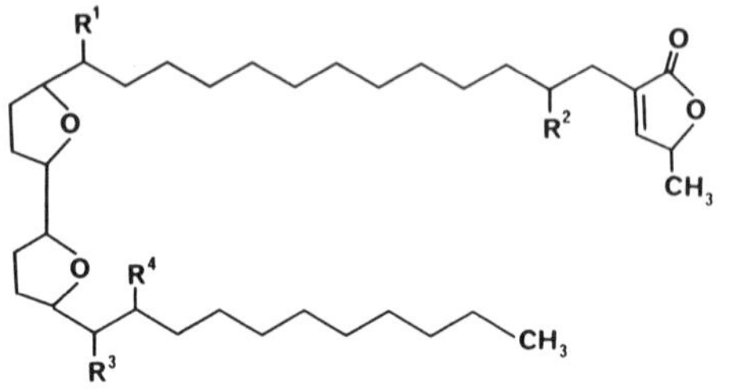

326 Ivarocom $R^1=OH$, $R^2=R^4=H$, $R^3=OCOCH_3$

327 Rollinicin $R^1=R^3=R^4=OH$, $R^2=H$

328 Isorollinicin $R^1=R^3=R^4=OH$, $R^2=H$

329 Asimicin $R^1=R^2=R^3=OH$, $R^4=H$

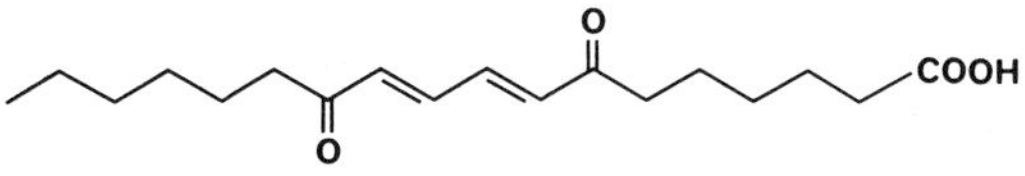

330 Ostopanic acid

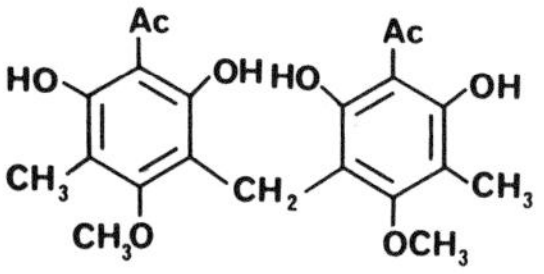

331 Mallotophenone

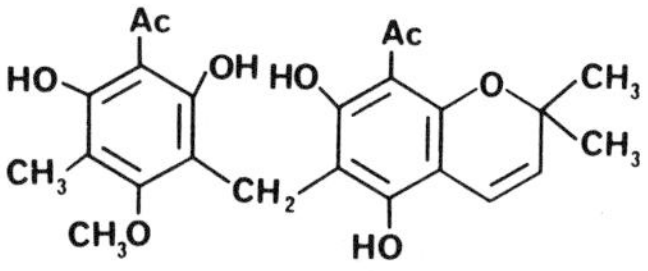

332 Mallotochromene

Cleomiscosin A (**333**) has been isolated from *Cleome viscosa* L. (Ray *et al.*, 1980, 1982), *Soulamea soulameoides* (Gray) Nooteboom, *Simaba multiflora* A. Juss, and *Matayaba arborescens* Radlk. (Arisawa *et al.*, 1984a). Cleomiscosin A (**333**) was inactive in the KB test system but exerted moderate activity in the P388 lymphocytic leukemia in cell culture. Cleomiscosin A (**333**) and its isomer cleomiscosin B (**334**) have been synthesized from the dihydroxycoumarin fraxetin (**335**) and coniferyl alcohol (**336**) by chemical or enzymic oxidation. In a similar manner, propacin (**337**) has been synthesized from fraxetin (**335**) and isoeugenol, and coumarolignan aquillochin (**338**) from fraxetin (**335**) and sinapyl alcohol (**339**). Starting from daphnetin (**340**) and sinapyl alcohol (**339**), daphneticin (**341**) has been obtained (Lin and Cordell, 1984). An efficient technique for the structure elucidation of coumarolignans has been developed by application of the SINEPT pulse program (Lin and Cordell, 1986); this resulted in the unambiguous structure determination of **333**, **334**, **337**, **338**, and a revision in the structure of **341**.

Withaferin A (**342**) and six analogs were isolated from the aerial parts of *Withania frutescens* Pauq. and *Withania aristata* Pauq. (Gonzales *et al.*, 1982). Withaferin A (**342**) and its 5-hydroxy-6-chloro derivative exhibited marked cytostatic activity against HeLa **229** cells in culture. The lactone moiety would appear to be a prerequisite for activity.

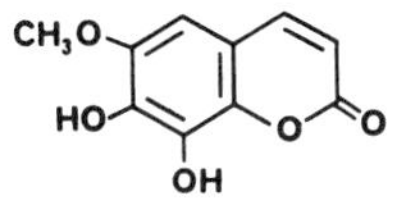

335 Fraxetin

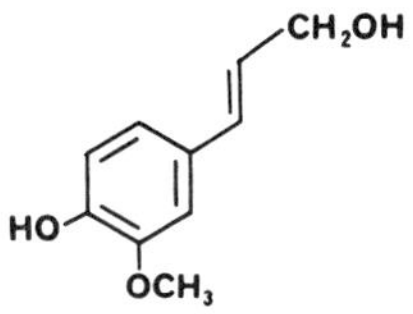

336 Coniferyl alcohol

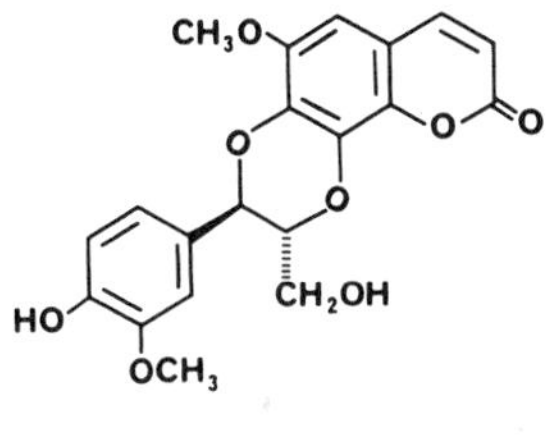

333 Cleomiscosin A

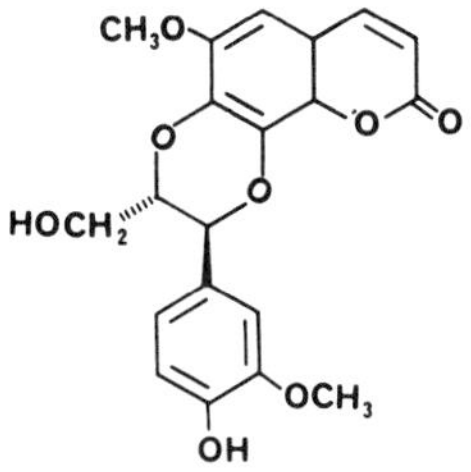

334 **Cleomiscosin B**

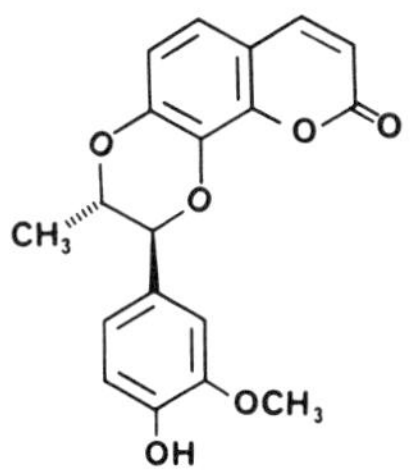

337 Propacin

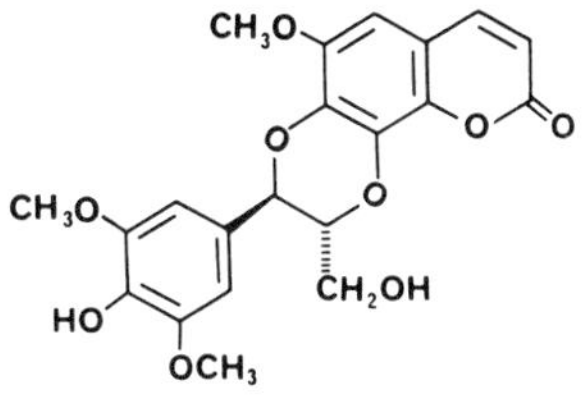

338 Aquillochin

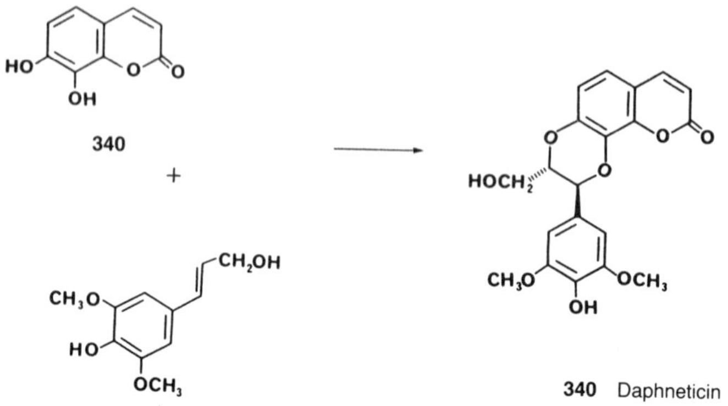

340 + 339 ⟶ 340 Daphneticin

A new cucurbitacin derivative, hexanorcucurbitacin F (**343**), has been isolated from *Elaeocarpus dolichostylus* Schltr. along with the known cucurbitacin F (**344**). The latter compound displayed significant cytotoxic activity in the KB and P388 *in vitro* cell culture test systems; however, **343** proved to be inactive in those assays (Fang *et al.*, 1984). Cucurbitacin I (**345**) has been isolated from the leaves of *Gyrinops walla* Gaertn. (Schun and Cordell, 1985). This was the first isolation of a cucurbitacin from a species of plant family Thymelaceaceae. Trewinine (**346**), a new cytotoxic cucurbitacin, has been isolated from *Trewia nudiflora* L. (Saha *et al.*, 1981). The structure of cucurbitacin S, isolated from *Bryonia dioica* Jacq., has been revised (Hylands and Mansour, 1982).

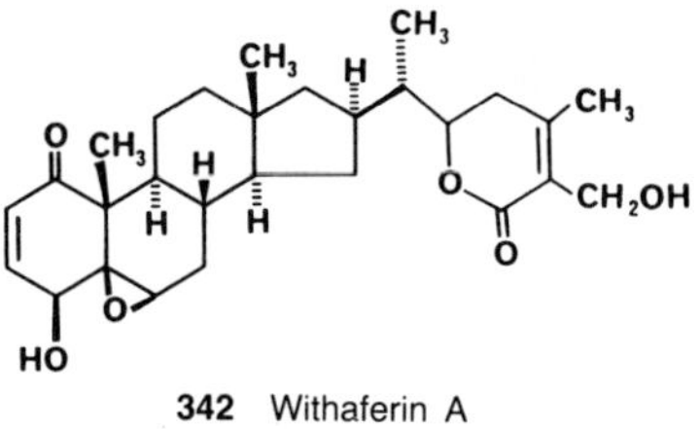

342 Withaferin A

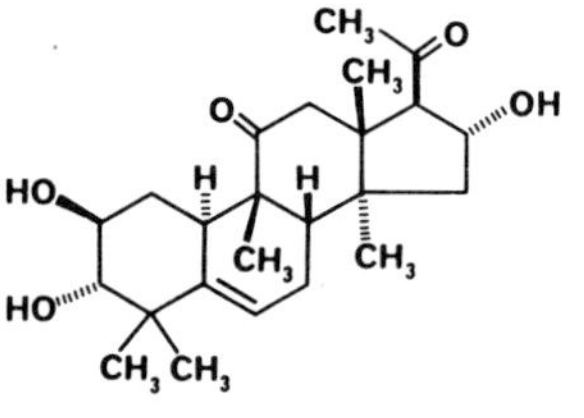

343 Hexanorcucurbitacin F

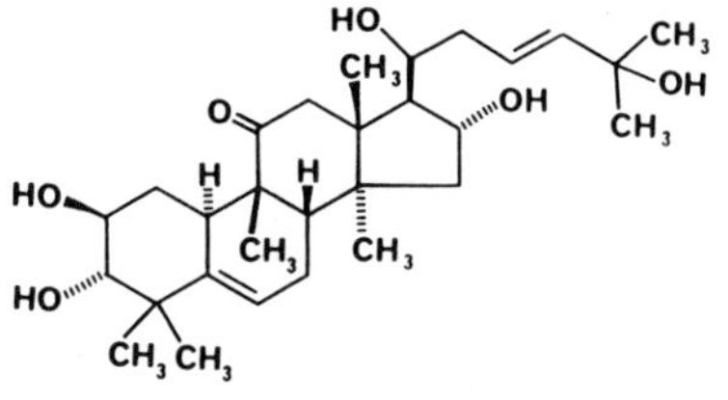

344 Cucurbitacin F

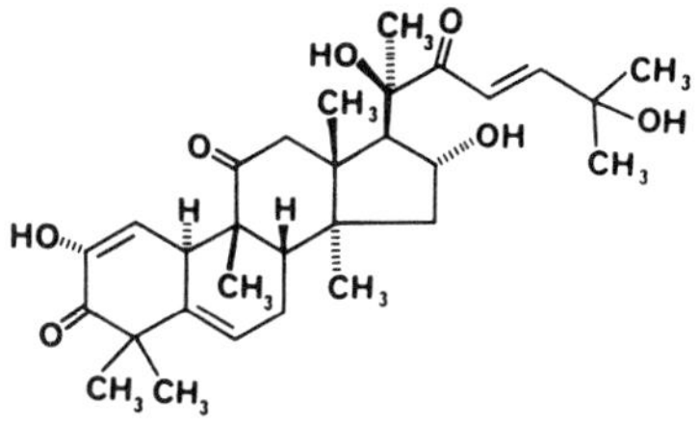

345 Cucurbitacin I

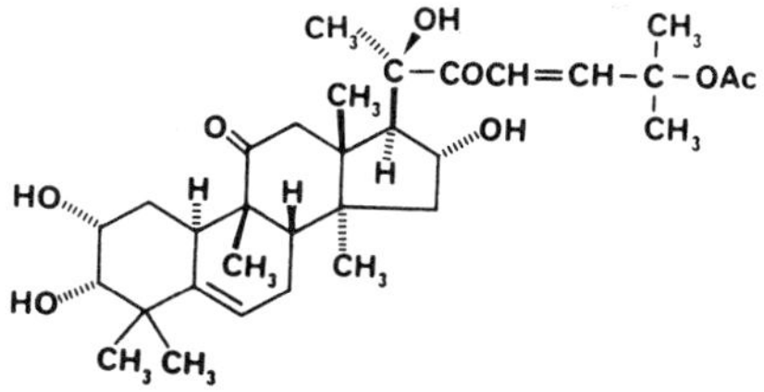

346 Trewinine

The bioassay-guided fractionation of the roots and leaves of *Cracaena afro-montana* Mildbraed yielded afromontoside (**347**), a new cytotoxic steroidal saponin (Reddy *et al.*, 1984). Separation of the constituents of *Dregea volubilis* (L.) Benth. afforded seven new glycosides, among them dregeoside A_{P1} and dregeoside A_{O1}, which were active against the solid-type Ehrlich carcinoma and B16 melanoma (Yoshimura *et al.*, 1983). Cytotoxic cucurbitacin glycosides, including datiscosides B(8), C(3), O(9), E(4), F(5), G(6), and H(10), have been isolated from the dried twigs of *Datisca glomerata* Baill. (Sasamori *et al.*, 1983).

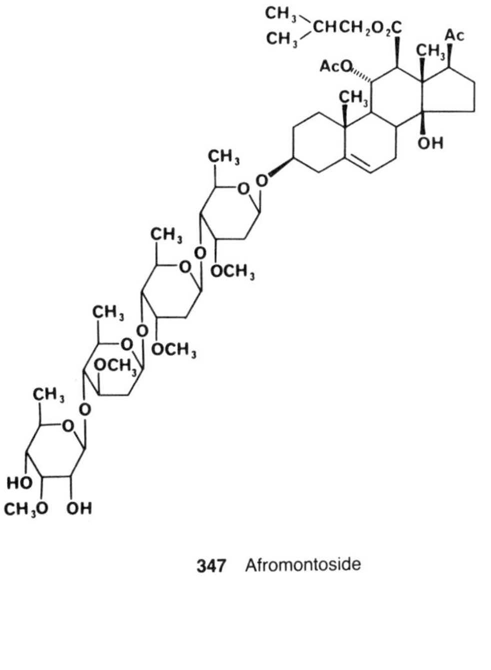

347 Afromontoside

III. SUMMARY

Plant anticancer agents continue to be a topic of substantial academic and commercial interest. Moreover, as this brief review amply demonstrates, success in this field continues to be demonstrated through the isolation of new biologically active compounds, through progress in the synthesis of various skeleta and individual active compounds, and through the use of new *in vitro* test systems. It can reasonably be assumed that as more diverse plants are examined, novel agents will continue to be discovered to stimulate the interest of the synthetic organic chemists and the biologists.

REFERENCES

Adachi, K., Ishikawa, M., Nakamura, K., Itoh, S., Saitoh, K., Yamashita, T., Takahira, T., Ozawa, Y., Sakitama, K., *et al.* (1985). *Oyo Yakuri* **30**, 527; *Chem. Abstr.* **104**, 28521k.

Arisawa, M., Fujita, A., Suzuki, R., Hayashi, T., Morita, N., Kawano, N., and Koshimura, S. (1985). *J. Nat. Prod.* **48**, 455.

Arisawa, M., Handa, S. S., McPherson, D. D., Lankin, D. C., Cordell, G. A., Fong, H. H. S., and Farnsworth, N. R. (1984a). *J. Nat. Prod.* **47**, 300.

Arisawa, M., Pezzuto, J. M., Bevelle, C., and Cordell, G. A. (1984b). *J. Nat. Prod.* **47**, 453.

Asada, Y., and Furuya, T. (1984). *Chem. Pharm. Bull.* **32**, 4616.

Atta-ur-Rahman (1980). Proceedings of 12th International IUPAC Symposium on Chemistry of Natural Products, Tenerife, p. 245.

Auclair, C., Dugne, B., Meunier, B., and Paoletti, C. (1986). *Biochemistry* **25**, 1240.

Auclair, C., Meunier, B., and Paoletti, C. (1983). *Biochem. Pharmacol.* **32**, 3883.

Auclair, C., Voisin, E., Banoun, H., Paoletti, C., Bernadou, J., and Meunier, B. (1984). *J. Med. Chem.* **27**, 1161.

Balkandzhieva, Yu. (1985). *Priroda (Sofia)* **34**, 32; *Chem. Abstr.* **104**, 141529g.

Batsuren, D., Batirov, Kh.E., Malikov, V. M., Zemlyanskii, V. N., and Yagudaeu, M. R. (1981). *Khim. Prir. Soedin.* **17**, 295.

Begley, M. J., Jackson, C. B., and Pattenden, G. (1985). *Tetrahedron Lett.* **26**, 3397.

Bertrand, J. R., and Giacomoni, P. V. (1985). *Chemioterapia* **4**, 445.

Bhatnagar, S., Polonsky, J., Prange, T., and Pascard, C. (1984). *Tetrahedron Lett.* **25**, 299.

Bhatnagar, S., Polonsky, J., Sevenet, T., and Prange, T. (1985). *Tetrahedron Lett.* **26**, 1225.

Bhushana, R. K. S. P., Collard, M.-P. M., Dejonghe, J.-P. C., Attassi, G., Hannart, J. A., and Trouet, A. (1985a). *J. Med. Chem.* **28**, 1079.

Bhushana, R. K. S. P., Collard, M.-P. M., and Trouet, A. (1985b). *Anticancer Res.* **5**, 379.

Bhutani, K. K., Ali, M., and Atal, C. K. (1984). *Phytochemistry* **23**, 1765.

Bowden, B. F., Read, R. W., and Taylor, W. C. (1980). *Aust. J. Chem.* **33**, 1823.

Boxenbaum, H., and Fertig, J. B. (1984). *Biopharm. Drug Dispos.* **5**, 405.

Brade, W. P. (1981). *Beitr. Onkol.* **6**, 95; *Chem. Abstr.* **95**, 143686a.

Bredenkamp, M. W., Wiechers, A., and van Rooyen, P. H. (1985). *Tetrahedron Lett.* **26**, 929.

Broggini, M., Colombo, T., and D'Incalci, M. (1982). *Cancer Treat. Rep.* **67**, 555.

Brossi, A., and Kerekes, P. (1984). U.S. Pat. Appl. 601,314; *Chem. Abstr.* **103**, 6576j.

Brossi, A., Teitel, S., and Parry, G. V. (1971). *In* "The Alkaloids, Vol. XIII" (R. H. F. Manske and H. L. Holmes, eds.), pp. 189–212. Academic Press, New York.

Brown, E., Loriot, M., and Robin, J.-P. (1982). *Tetrahedron Lett.* **23**, 949.

Burton, M., and Robins, D. J. (1985). *J. Chem. Soc., Perkin Trans.* **1**, 611.

Cai, J.-C., and Hutchinson, C. R. (1983a). *Chem. Heterocycl. Compd.* **25**, 753.

Cai, J.-C., and Hutchinson, C. R. (1983b). *In* "The Alkaloids, Vol. XXI" (A. Brossi, ed.), pp. 101–138. Academic Press, New York.

Capraro, H. G., and Brossi, A. (1984). *In* "The Alkaloids, Vol. XXIII" (A. Brossi, ed.), pp. 1–70. Academic Press, New York.

Cassady, J. M., Chang, C.-J., and McLaughlin, J. L. (1981). *In* "Natural Products as Medical Agents" (J. L. Beal and E. Reinhard, eds.), p. 93. Hippocrates Verlag, Stuttgart.

Cassady, J. M., and Douros, J. D. (eds.) (1980). "Anticancer Agents Based on Natural Products Models". Academic Press, New York.

Cassady, J. M., and Suffness, M. (1980). *In* "Anticancer Agents Based on Natural Product Models" (J. M. Cassady, and J. D. Douros, eds.), p. 201. Academic Press, New York.

Chamberlin, A. R., and Chung, J. Y. L. (1983). *J. Am. Chem. Soc.* **105**, 3653.

Chamberlin, A. R., and Chung, J. Y. L. (1985). *J. Org. Chem.* **50**, 4425.

Charcosset, J. Y., Jacquemin-Sablon, A., and Le Pecq, J. B. (1984). *Biochem. Pharmacol.* **33**, 2271.

Charles, B., Bruneton, J., and Cave, A. (1986). *J. Nat. Prod.* **49**, 303.

Cheng, J., Zhang, J., Zhang, Q., Yang, J., and Huang, L. (1984). *Yaoxue Xuebao* **19**, 178; *Chem. Abstr.* **103**, 160738v.

Chommadov, B., Yusupov, M. K., and Aslanov, Kh. A. (1985). *Khim. Prir. Soedin.* **21**, 417.

Cordell, G. A. (1977). *In* "New Natural Products and Plant Drugs with Pharmacological, Biological or Therapeutical Activity" (H. Wagner and P. Wolff, eds.), pp. 54–81. Springer-Verlag, Berlin.

Cordell, G. A. (1979). *In* "The Alkaloids, Vol. XVII" (R. H. F. Manske, and R. G. A. Rodrigo, eds.), pp. 199–384. Academic Press, New York.

Cordell, G. A. (1983). *In* "Monoterpenoid Indole Alkaloids" (J. E. Saxton, ed.), pp. 539–727. John Wiley and Sons, Inc., New York.

Cordell, G. A., and Saxton, J. E. (1981). *In* "The Alkaloids, Vol. XX" (A. Brossi, ed.), pp. 1–295. Academic Press, New York.

Creaven, P. J. (1982). *Curr. Drugs Methods Cancer Treat.*, p. 61; *Chem. Abstr.* **100**, 131888b.

Creaven, P. J. (1983). *Etoposide (VP-16): Curr. Status New Dev.* [Pap. Symp.] p. 103; *Chem. Abstr.* **101**, 122373y.

Cros, S., and Raynaud, A. (1984). *Arch. Biol.* **95**, 113.

Cui, Z., Qi, M., Lin, L., and Yu, D. (1984). *Yaoxue Xuebao* **19**, 904; *Chem Abstr.* **103**, 19853x.

Cushman, M., and Chen, J.-K. (1986). Proceedings of the 27th Annual Meeting of the American Society of Pharmacognosy, Ann Arbor, Michigan.

Cushman, M., and Mohan, P. (1985). *J. Med. Chem.* **28**, 1031.

Dabrah, T. T., and Sneden, A. T. (1984). *Phytochemistry* **23**, 2013.

Dalvi, R. R. (1985). *Experientia* **41**, 77.

Das, K. G., Afzal, J., Hazra, B. G., and Bhawal, B. M. (1983). *Synth. Commun.* **13**, 787.

De Bruyn, A., De Taeye, L., Simonds, R., Verzele, M., and De Pauw, C. (1982). *Bull. Soc. Chim. Belg.* **91**, 75.

Delbarre, A., Oberlin, R., Rouges, B. P., Borgna, J.-L., Rochefort, H., Le Pecq, J.-B., and Jacquemin-Sablon, A. (1985). *J. Med. Chem.* **28**, 752.

Dewick, P. M., and Jackson, D. E. (1981). *Phytochemistry* **20**, 2277.

Dhal, R., Brown, E., and Robin, J.-P. (1983). *Tetrahedron* **39**, 2787.

Differding, E., and Ghosez, L. (1985). *Tetrahedron Lett.* **26**, 1647.

D'Incalci, M., Farina, P., Sessa, C., Mangioni, C., Conter, V., Masera, G., Rocchetti, M., Brambilla, P. M., Piazza, E., Beer, M., and Cavalli, F. (1982). *Cancer Chemother. Pharmacol.* **7**, 141.

Dvorackova, S., Guenard, D., Picot, F., Simanek, V., and Waisser, K. (1985). *Acta Univ. Palacki. Olomuc., Fac. Med.* **111**, 13; *Chem. Abstr.* **104**, 141708g.

El-Sayed, A., and Cordell, G. A. (1981). *J. Nat. Prod.* **44**, 289.

El-Sayed, A., Handy, G. A., and Cordell, G. A. (1980). *J. Nat. Prod.* **43**, 157.

El-Sayed, A., Handy, G. A., and Cordell, G. A. (1983). *J. Nat. Prod.* **46**, 517.

Evidente, A. (1986). *J. Nat. Prod.* **49**, 90.

Faddejeva, M. D., Belyaeva, T. N., Rosanov, Yu. M., Sedova, V. M., and Sokolovskaya, E. L. (1984). *Stud. Biophys.* **104**, 267.

Fang, X.-D., Phoebe, C. H., Pezzuto, J. M., Fong, H. H. S., Farnsworth, N. R., Yellin, B., and Hecht, S. M. (1984). *J. Nat. Prod.* **47**, 988.

Findlay, J. A. (1976). *In* "Alkaloids, International Review of Science Series Two, Organic Chemistry, Vol. 9" (K. Wiesner, ed.), p. 23. Butterworths, London.

Fonesca, S. F., Ruveda, E. A., and McChesney, J. D. (1980). *Phytochemistry* **19**, 1527.

Forgacs, P., Provost, J., Fuche, A., Guenard, D., Thal, C., and Guilhem, J. (1985). *Tetrahedron Lett.* **26**, 3457.

Fuganti, C. (1975). *In* "The Alkaloids, Vol. XV" (R. H. F. Manske and H. L. Holmes, eds.), pp. 83–164. Academic Press, New York.

Fujii, T., Kogen, H., Yoshifuji, S., and Ohba, M. (1985). *Chem. Pharm. Bull.* **33**, 1946.

Fujii, T., and Ohba, M. (1983). *In* "The Alkaloids, Vol. XXII" (A. Brossi, ed.), pp. 1–50. Academic Press, New York.

Fujii, T., and Ohba, M. (1985a). *Chem. Pharm. Bull.* **33**, 144.

Fujii, T., and Ohba, M. (1985b). *Chem. Pharm. Bull.* **33**, 583.

Fukada, M. (1985). *Biochem. Pharmacol.* **34**, 1225.

Funayama, S., Borris, R. P., and Cordell, G. A. (1983). *J. Nat. Prod.* **46**, 391.

Funayama, S., and Cordell, G. A. (1983). *Planta Med.* **48**, 263.

Funayama, S., and Cordell, G. A. (1984). *Planta Med.* **49**, 117.

Funayama, S., and Cordell, G. A. (1985a). *J. Nat. Prod.* **48**, 114.

Funayama, S., and Cordell, G. A. (1985b). *J. Nat. Prod.* **48**, 938.

Funayama, S., and Cordell, G. A. (1986). *J. Nat. Prod.* **49**, 210.

Funayama, S., Cordell, G. A., Macfarlane, R. D., and McNeal, C. J. (1985). *J. Org. Chem.* **50**, 1737.

Funayama, S., Cordell, G. A., Wagner, H., and Lotter, H. L. (1984). *J. Nat. Prod.* **47**, 143.

Furuno, T., Ishibashi, M., Naora, H., Murae, T., Hirota, H., Tsujiki, T., Takahashi, T., Itai, A., and Iitaka, Y. (1984). *Bull. Chem. Soc. Jpn.* **57**, 2484.

Gensler, W. J., and Gatsonis, C. D. (1966). *J. Org. Chem.* **31**, 4004.

Gero, M. R., Bachrach, U., Bhatnagar, S., and Polonsky, J. (1985). *C.R. Acad. Sci. Ser. 2.* **300**, 803.

Gerzon, K., and Svoboda, G. H. (1983). *In* "The Alkaloids, Vol. XXI" (A. Brossi, ed.), pp. 1–29. Academic Press, New York.

Ghosal, S., Kumar, Y., Singh, S. K., and Kumar, A. (1986). *J. Chem. Res. (S)* p. 112.

Ghosal, S., Lochan, R., Ashutosh, Kumar, Y., and Srivastava, R. S. (1985b). *Phytochemistry* **24**, 1825.

Ghosal, S., Saini, K. S., Razdan, S., and Kumar, Y. (1985c). *J. Chem. Res. (S)* p. 100.

Ghosal, S., Shanthy, A., Kumar, A., and Kumar, Y. (1985a). *Phytochemistry* **24**, 2703.

Glinski, M. B., and Durst, T. (1983). *Can. J. Chem.* **61**, 573.

Gonzales, A. G., Darias, V., Martin Herrera, D. A., and Suarez, M. S. (1982). *Fitoterapia* **53**, 85.

Gottlieb, H. E., Mervic, M., Ghera, E., and Frolow, F. (1982). *J. Chem. Soc., Perkin Trans.* **1**, 2353.
Gouyette, A., Auclair, C., and Paoletti, C. (1985). *Biochem. Biophys. Res. Commun.* **131**, 619.
Gribble, G. W. (1985). *Heterocycles* **23**, 1277.
Grieco, P. A., Ferrino, S., and Vidari, G. (1980a). *J. Am. Chem. Soc.* **102**, 7586.
Grieco, P. A., Lis, R., Ferrino, S., and Jaw, J. Y. (1982). *J. Org. Chem.* **47**, 601.
Grieco, P. A., Lis, R., Ferrino, S., and Jaw, J. Y. (1984a). *J. Org. Chem.* **49**, 2342.
Grieco, P. A., Oguri, T., and Gilman, S. (1980b). *J. Am. Chem. Soc.* **102**, 5886.
Grieco, P. A., Skam, H.-L., Inanga, J., Kim, H., and Tuthill, P. A. (1984b). *J. Chem. Soc., Chem. Commun.* p. 1345.
Grundon, M. F. (1984). *Nat. Prod. Rep.* **1**, 247.
Grundon, M. F. (1985a). *Nat. Prod. Rep.* **2**, 235.
Grundon, M. F. (1985b). *Nat. Prod. Rep.* **2**, 249.
Gueritte, F., Langlois, N., Langlois, Y., Sundberg, R. J., and Bloom, J. D. (1981). *J. Org. Chem.* **46**, 5393.
Gunawardana, Y. A. G. P., and Cordell, G. A. (1987). Unpublished results.
Hagglund, K. M., L'Emperur, K. M., Roby, M. R., and Stermitz, F. R. (1985). *J. Nat. Prod.* **48**, 638.
Hall, I. H., Liou, Y. F., Lee, K.-H., Chaney, S. G., and Willingham, W. (1983). *J. Pharm. Sci.* **72**, 626.
Hamburger, M., Handa, S. S., Cordell, G. A., Kinghorn, A. D., and Farnsworth, N. R. (1987). *J. Nat. Prod.* **50**, 19.
Hanaoka, M., Yamagishi, H., and Mukai, C. (1985b). *Chem. Pharm. Bull.* **33**, 1763.
Hanaoka, M., Yoshida, S., and Mukai, C. (1984). *J. Chem. Soc., Chem. Commun.* p. 1703.
Hanaoka, M., Yoshida, S., and Mukai, C. (1985a). *Tetrahedron Lett.* **26**, 5163.
Hayes, M. A., Roberts, E., and Farber, E. (1985). *Cancer Res.* **45**, 3726.
Heathcock, C. H., Mahaim, C., Schlecht, M. F., and Utawanit, T. (1984). *J. Org. Chem.* **49**, 3264.
Hewlins, M. J. E., Oliveira-Campos, A.-M., and Shannon, P. V. R. (1984). *Synthesis* **14**, 289.
Hicks, R. P., and Sneden, A. T. (1985). *J. Nat. Prod.* **48**, 357.
Hirschmann, S., Banerjee, G., and Jakupovic, S. (1985). *Rev. Latinoam. Quim.* **16**, 109; *Chem. Abstr.* **104**, 126516g.
Honda, T., Murae, T., Ohta, S., Kurata, Y., Kawai, H., Takahashi, T., Itai, A., and Iitaka, Y. (1981). *Chem. Lett.* p. 299.
Houlbert, N., Brown, E., Robin, J.-P., Davoult, D., Chiaroni, A., Prange, T., and Riche, C. (1985). *J. Nat. Prod.* **48**, 345.
Hsiang, Y. H., Hertzberg, R., Hecht, S., and Liu, L. F. (1985). *J. Biol. Chem.* **260**, 14873.
Huang, L., and Xue, Z. (1984). *In* "The Alkaloids, Vol. XXIII" (A. Brossi, ed.), pp. 157–226. Academic Press, New York.
Hudlicky, T., Frazier, J. O., and Kwart, L. D. (1985). *Tetrahedron Lett.* **26**, 3523.
Hudlicky, T., Frazier, J. O., Seoane, G., Tiedje, M., Seoane, A., Kwart, L. D., and Beal, C. (1986). *J. Am. Chem. Soc.* **108**, 3755.
Hulbert, P. B., Klyne, W., and Scopes, P. M. (1981). *J. Chem. Res. (S)* p. 27.
Hutchinson, C. R. (1981). *Tetrahedron* **37**, 1047.
Hylands, P. J., and Mansour, E. S. (1982). *Phytochemistry* **21**, 2703.
Iida, H., Watanabe, Y., and Kibayashi, C. (1985). *J. Chem. Soc., Perkin Trans.* **1**, 261.
Iida, H., Watanabe, Y., Tanaka, M., and Kibayashi, C. (1984). *J. Org. Chem.* **49**, 2412.
Ikeya, Y., Ookawa, N., Taguchi, H., and Yosioka, I. (1982a). *Chem. Pharm. Bull.* **30**, 3202.
Ikeya, Y., Taguchi, H., Sasaki H., Nakajima, K., and Yosioka, I. (1980a). *Chem. Pharm. Bull.* **28**, 2414.

Ikeya, Y., Taguchi, H., and Yosioka, I. (1978). *Chem. Pharm. Bull.* **26**, 682.
Ikeya, Y., Taguchi, H., and Yosioka, I. (1979a). *Chem. Pharm. Bull.* **27**, 2536.
Ikeya, Y., Taguchi, H., and Yosioka, I. (1980b). *Chem. Pharm. Bull.* **28**, 2422.
Ikeya, Y., Taguchi, H., and Yosioka, I. (1982b). *Chem. Pharm. Bull.* **30**, 3207.
Ikeya, Y., Taguchi, H., Yosioka, I., Iitaka, Y., and Kobayashi, H. (1979b). *Chem. Pharm. Bull.* **27**, 1395.
Ikeya, Y., Taguchi, H., Yosioka, I., and Kobayashi, H. (1979c). *Chem. Pharm. Bull.* **27**, 1576.
Ikeya, Y., Taguchi, H., Yosioka, I., and Kobayashi, H. (1979d). *Chem. Pharm. Bull.* **27**, 1583.
Ikeya, Y., Taguchi, H., Yosioka, I., and Kobayashi, H. (1979e). *Chem. Pharm. Bull.* **27**, 1383.
Ikeya, Y., Taguchi, H., Yosioka, I., and Kobayashi, H. (1979f). *Chem. Pharm. Bull.* **27**, 2695.
Ikeya, Y., Taguchi, H., Yosioka, I., and Kobayashi, H. (1980c). *Chem. Pharm. Bull.* **28**, 3357.
Inanga, J., Sasaka, S., Grieco, P. A., and Kim, H. (1985). *J. Am. Chem. Soc.* **107**, 4800.
Ishibashi, H., Ozeki, H., and Ikeda, M. (1986). *J. Chem. Soc., Chem. Commun.* p. 654.
Ishikawa, H. (1984). *Seitai No Kagaku* **35**, 518; *Chem. Abstr.* **102**, 197342s.
Issell, B. F., Tihon, C., and Curry, M. E. (1982). *Cancer Chemother. Pharmacol.* **7**, 113.
Jackson, D. V., Long, T. R., Trahey, T. F., and Morgan, T. M. (1984). *Cancer Chemother. Pharmacol.* **13**, 176.
Jackson, C. B., and Pattenden, G. (1985). *Tetrahedron Lett.* **26**, 3393.
Jatztold-Howorko, R., Bisagni, E., and Chermann, J. C. (1984). *Eur. J. Med. Chem.-Chim. Ther.* **19**, 541.
Jolad, S. D., Hoffman, J. J., Schram, K. H., Cole, J. R., Tempesta, M. S., Kriek, G. R., and Bates, R. B. (1982). *J. Org. Chem.* **47**, 3151.
Jolad, S. D., Wiedhoff, R. M., and Cole, J. R. (1975). *J. Pharm. Sci.* **64**, 1889.
Jordan, M. A., Himes, R. H., and Wilson, L. (1985). *Cancer Res.* **45**, 2741.
Jovanovics, K., Gorog, S., Eckhard, S., Sugar, J., Somfai, Zs., Farkas, E., and Hindy, I. (1980). Eur. Pat. Appl. 18,321; *Chem. Abstr.* **95**, 25380y.
Kanai, K., Zelle, R. E., Sham, H.-L., Grieco, P. A., and Callant, P. (1984). *J. Org. Chem.* **49**, 3867.
Kan-Fan, C., Freire, R., Husson, H. P., Fujii, T., and Ohba, M. (1985). *Heterocycles* **23**, 1089.
Karlsson, B., Pilotti, A. M., Soderholm, A. C., Norin, T., Sundin, S., and Sumimoto, M. (1978). *Tetrahedron* **34**, 234.
Kato, T., Takamoto, S., Mizutani, M., Hato, M., and Ota, K. (1984). *Gan To Kagaku Ryoho* **11**, 2393.
Kedzierski, B., and Buhler, D. R. (1985). *Toxicol. Lett.* **25**, 115.
Kedzierski, B., and Buhler, D. R. (1986a). *Anal. Biochem.* **152**, 59.
Kedzierski, B., and Buhler, D. R. (1986b). *Chem.-Biol. Interact.* **57**, 217.
Kende, A. S., Johnson, S., Sanfilippo, P., Hodges, J. C., and Jungheim, L. N. (1986). *J. Am. Chem. Soc.* **108**, 3513.
Kende, A. S., Liebeskind, L. S., Mills, J. E., Rutledge, P. S., and Curran, D. P. (1977). *J. Am. Chem. Soc.* **99**, 6082.
Kerekes, P., Brossi, A., Flippen-Anderson, J. L., and Chignell, C. F. (1985b). *Helv. Chim. Acta* **68**, 571.
Kerekes, P., Sharma, P. N., Brossi, A., Chignell, C. F., and Quinn, F. R. (1985a). *J. Med. Chem.* **28**, 1204.
Ketcha, D. M., and Gribble, G. W. (1985). *J. Org. Chem.* **50**, 5451.
Khalid, S. A., and Waterman, P. G. (1985). *J. Nat. Prod.* **48**, 118.
Krane, B. D., Fagbule, M. O., and Shamma, M. (1984). *J. Nat. Prod.* **47**, 1.
Kreis, W., Budman, D. R., Schulman, P., Freeman, J., Greist, A., Nelson, R. L., Marks, M., and Kevill, L. (1986). *Cancer Chemother. Pharmacol.* **16**, 70.

Kupchan, S. M., Britton, R. W., Ziegler, M. F., Gilmore, C. J., Restivo, R. J. and Bryan, R. F. (1973). *J. Am. Chem. Soc.* **95**, 1335.
Kuroda, R., and Sainsbury, M. (1984). *J. Chem. Soc., Perkin Trans.* **2**, 1751.
Kutney, J. P. (1978). *Lect. Heterocycl. Chem.* **4**, 59.
Kuwahara, J., Suzuki, T., and Sugiura, Y. (1985). *Nucleic Acids Symp. Ser.* **16**, 201.
Lafranconi, W. M., Ohkuma, S., and Huxtable, R. J. (1985). *Toxicon* **23**, 983.
Lamberton, J. A. (1984). *Nat. Prod. Rep.* **1**, 245.
Langlois, N., Andriamialisova, R. Z., and Neuss, N. (1980). *Helv. Chim. Acta* **63**, 793.
Langlois, N., Gueritte, F., Langlois, Y., and Potier, P. (1976). *J. Am. Chem. Soc.* **98**, 7017.
Larson, E. R., and Raphael, R. A. (1982). *J. Chem. Soc., Perkin Trans.* **1**, 521.
Lathbury, D., and Gallagher, T. (1986). *J. Chem. Soc., Chem. Commun.* p. 1017.
Lei, H. (1982). *Yaoxue Xuebao* **17**, 1; *Chem. Abstr.* **96**, 193445n.
Li, Y., Li, Y., and Huang, W. (1984b). *Kexue Tongbao* **29**, 1131; *Chem. Abstr.* **102**, 122319v.
Li, Y., Wu, K., and Huang, L. (1984a). *Yaoxue Xuebao* **19**, 582; *Chem. Abstr.* **102**, 132308r.
Li, Z., and Han, R. (1986). *Zhongcaoyao* **17**, 135; *Chem. Abstr.* **105**, 17632s.
Lin, L.-J., and Cordell, G. A. (1984). *J. Chem. Soc., Chem. Commun.* p. 160.
Lin, L.-J., and Cordell, G. A. (1986). *J. Chem. Soc., Chem. Commun.* p. 377.
Lin, W., Chen, R., and Xue, Z. (1985). *Yaoxue Xuebao* **20**, 283; *Chem. Abstr.* **103**, 85049r.
Loike, J. D. (1984). *Trends Pharmacol. Sci.* **5**, 30.
Long, B. H., and Brattain, M. G. (1983). *Etoposide (VP-16): Curr. Status New Dev.* [Pap. Symp.] p. 63; *Chem. Abstr.* **101**, 163287.
Long, B. H., Musial, S. T., and Brattain, M. G. (1985). *Cancer Res.* **45**, 3106.
Lumonadio, L., and Vanhaelen, M. (1985). *Phytochemistry* **24**, 2387.
Luyckx, M., Brunet, C., and Cazin, M. (1984). *Methods Find. Exp. Clin. Pharmacol.* **6**, 679.
Mackay, M. F., Mitrprachachon, P., Oliver, P. J., and Culvenor, C. C. J. (1985). *Acta Crystallogr., Sect. C: Cryst. Struct. Commun.* **C41**, 722.
Madhusudanan, K. P., Gupta, S., and Bhakuni, D. S. (1984). *Indian J. Chem., Sect. B.* **23B**, 1012.
Maftouh, M., Amiar, Y., and Picard-Fraire, C. (1985b). *Biochem. Pharmacol.* **34**, 427.
Maftouh, M., Besselievre, R., Monsarrat, B., Lesca, P., Meunier, B., Husson, H. P., and Paoletti, C. (1985a). *J. Med. Chem.* **28**, 708.
Magnus, P., Schultz, J., and Gallagher, T. (1984). *J. Chem. Soc., Chem. Commun.* p. 1179.
Magnus, P., Schultz, J., and Gallagher, T. (1985). *J. Am. Chem. Soc.* **107**, 4984.
Magri, N. F., and Kingston, D. G. I. (1986). *J. Org. Chem.* **51**, 797.
Malvy, C., Prevost, P., Gansser, C., Viel, C., and Paoletti, C. (1986). *Chem.-Biol. Interact.* **57**, 41.
Mann, J., and Piper, S. E. (1982). *J. Chem. Soc., Chem. Commun.* p. 430.
Maral, R., Bourot, C., Chenu, E., and Mathe, G. (1984). *Cancer Lett.* **22**, 49.
Meyers, A. I., and Avila, W. B. (1981). *J. Org. Chem.* **46**, 3881.
Miller, R. W. (1980). *J. Nat. Prod.* **43**, 425.
Mitchell, R. E., Stocklin, W., Stefanovic, M., and Geissman, T. A. (1971). *Phytochemistry* **10**, 411.
Miyaji, K., Nakamura, T., Hirota, H., Igarashi, M., and Takahashi, T. (1984). *Tetrahedron Lett.* **25**, 5299.
Mori, H., Sugie, S., Yoshimi, N., Asada, Y., Furuya, T., and Williams, G. M. (1985). *Cancer Res.* **45**, 3125.
Morita, M., Haji, A., Goto, A., Hattori, N., and Hasegawa, Y. (1986). *Nippon Yakugaku Zasshi* **87**, 53; *Chem. Abstr.* **104**, 122710t.
Morufushi, H. (1984). *Seitai No Kagaku* **35**, 524; *Chem. Abstr.* **102**, 180265w.
Mukhopadhyay, S., and Cordell, G. A. (1981). *J. Nat. Prod.* **44**, 611.

Mulchandani, N. B., and Venkatachalam, S. R. (1984). *Phytochemistry* **23**, 1206.
Mullin, K., Houghton, P. J., Houghton, J. A., and Horowitz, M. E. (1985). *Biochem. Pharmacol.* **34**, 1975.
Murphy, W. S., and Wattanasin, S. (1982). *J. Chem. Soc., Perkin Trans.* **1**, 271.
Murphy, W. S., and Wattanasin, S. (1980). *J. Chem. Soc., Chem. Commun.* p. 262.
Nandi, R., and Maiti, M. (1985). *Biochem. Pharmacol.* **34**, 321.
Newlands, E. S. (1983). *Dev. Oncol.* p. 15.
Ninomiya, I., and Naito, T. (1983). *In* "The Alkaloids, Vol. XXII" (A. Brossi, ed.), pp. 189–281. Academic Press, New York.
Nishimura, Y., Kondo, S., and Umezawa, H. (1985). *J. Org. Chem.* **50**, 5210.
Nishimura, Y., Kondo, S., and Umezawa, M. (1986). *Tetrahedron Lett.* **27**, 4323.
Niwa, H., Miyachi, Y., Okamoto, O., Uosaki, Y., and Yamada, K. (1986b). *Tetrahedron Lett.* **27**, 4605.
Niwa, H., Miyachi, Y., Uosaki, Y., Kuroda, A., Ishiwata, M., and Yamada, K. (1986c). *Tetrahedron Lett.* **27**, 4609.
Niwa, H., Miyachi, Y., Uosaki, Y., and Yamada, K. (1986a). *Tetrahedron Lett.* **27**, 4601.
Okano, M., Fujita, T., Fukamiya, N., and Aratani, T. (1984). *Chem. Lett.* p. 221.
Okano, M., Fujita, T., Fukamiya, N., and Aratani, T. (1985b). *Bull. Chem. Soc. Jpn.* **58**, 1793.
Okano, M., Fukamiya, N., Aratani, T., Juichi, M., and Lee, K.-H. (1985a). *J. Nat. Prod.* **48**, 972.
Olaniyi, A. A., and Powell, J. W. (1980). *J. Nat. Prod.* **43**, 482.
Ookawa, N., Ikeya, Y., Taguchi, H., and Yosioka, I. (1981). *Chem. Pharm. Bull.* **29**, 123.
Pakrashi, S. C., and Ali, E. (1967). *Tetrahedron Lett.* p. 2143.
Pearce, H. L. (1983). U.S. Patent 4,430,269; *Chem. Abstr.* **100**, 192142a.
Pearson, W. H. (1985). *Tetrahedron Lett.* **26**, 3527.
Pestchanker, M. J., Ascheri, M. S., and Giordano, O. S. (1985a). *Planta Med.* **51**, 165.
Pestchanker, M. J., Ascheri, M. S., and Giordano, O. S. (1985b). *Phytochemistry* **24**, 1622.
Pettit, G. R. (1977). "Biosynthetic Products for Cancer Chemotherapy Vol. I". Plenum Press, New York.
Pettit, G. R., and Cragg, G. M. (1978). "Biosynthetic Products for Cancer Chemotherapy Vol. II". Plenum Press, New York.
Pettit, G. R., and Ode, R. H. (1979). "Biosynthetic Products for Cancer Chemotherapy Vol. III". Plenum Press, New York.
Phillips, N. C., and Lauper, R. D. (1983). *Clin. Pharm.* **2**, 112.
Plaumann, H. P., Smith, J. G., and Rodrigo, R. (1980). *J. Chem. Soc., Chem. Commun.* p. 354.
Pokorny, E., Szikla, K., Palyi, I., and Holczinger, L. (1983). *Eur. J. Cancer Clin. Oncol.* **19**, 1113.
Polonsky, J. (1973). *Fortschr. Chem. Org. Naturst.* **30**, 101.
Polonsky, J. (1983). *Ann. Proc. Phytochem. Soc. Eur.* **22**, 247.
Polonsky, J. (1985). *Fortschr. Chem. Org. Naturst.* **47**, 221.
Polonsky, J., Bhatnagar, S., and Moretti, C. (1984). *J. Nat. Prod.* **47**, 994.
Potier, P. (1980a). *Ann. Pharm. Fr.* **38**, 407.
Potier, P. (1980b). *J. Nat. Prod.* **43**, 72.
Prasad, M. R. N., and Diszfalusy, E. (1982). *Int. J. Androl. Suppl.* **5**, 53.
Pratviel, G., Bernadou, J., Meunier, B., and Paoletti, C. (1985a). *Nucleosides Nucleotides* **4**, 143.
Pratviel, G., Bernadou, J., Paoletti, C., Meunier, B., Gillet, B., Guittet, E., and Lallemand, J. Y. (1985b). *Biochem. Biophys. Res. Commun.* **128**, 1173.
Qiu, C., and Wu, G. (1984). *Shengwu Huaxue Yu Sengwu Wuli Xuebao* **16**, 645; *Chem. Abstr.*

103, 16524e.

Rajapaksa, D., and Rodrigo, R. (1981). *J. Am. Chem. Soc.* **103**, 6208.

Rao, P. N., Wang, Y. C., Lotzova, E., Khan, A. A., Rao, S. P., and Stephens, L. C. (1985). *Cancer Chemother. Pharmacol.* **15**, 20.

Ray, A. B., Chattopadhyay, S. K., Konno, C., and Hikino, H. (1980). *Tetrahedron Lett.* **21**, 4477.

Ray, A. B., Chattopadhyay, S. K., Konno, C., and Hikino, H. (1982). *Heterocycles* **19**, 19.

Reddy, K. S., Shekhani, M. S., Berry, D. E., Lynn, D. G., and Hecht, S. M. (1984). *J. Chem. Soc., Perkin Trans.* **1**, 987.

Reider, P. J., and Roland, D. M. (1984). *In* "The Alkaloids, Vol. XXIII" (A. Brossi, ed.), pp. 71–156. Academic Press, New York.

Reisch, J., and Aly, S. (1986). *Arch. Pharm. (Weinheim)* **319**, 25.

Renz, J., Kuhn, M., and von Wartburg, A. (1965). *Liebigs Ann. Chem.* **681**, 207.

Rithner, C. D., Bushweller, C. H., Gensler, W. J., and Hoogasian, S. (1983). *J. Org. Chem.* **48**, 1491.

Robin, J.-P., Dhal, R., and Brown, E. (1982). *Tetrahedron* **38**, 3667.

Robins, D. J. (1982). *In* "Progress in the Chemistry of Organic Natural Products Vol. 41" (W. Herz, H. Grisebach and G. W. Kirby, eds.), p. 116. Springer-Verlag, Berlin.

Robins, D. J. (1984). *Nat. Prod. Rep.* **1**, 235.

Robins, D. J. (1985). *Nat. Prod. Rep.* **2**, 213.

Rodrigo, R. (1980). *J. Org. Chem.* **45**, 4538.

Rupprecht, J. K., Chang, C. J., Cassady, J. M., McLaughlin, J. L., Nikolajaczak, K. L., and Weisleder, D. (1986). *Heterocycles* **24**, 1197.

Saifah, E., Kelley, G. J., and Leany, J. D. (1983). *J. Nat. Prod.* **46**, 353.

Sainsbury, M. (1977). *Synthesis* **7**, 437.

Saha, P. K., Mukherjee, S., Shaw, A. K., Ganguly, T., and Ganguly, S. N. (1981). *Fitoterapia* **52**, 231.

Sakaki, T., Yoshimura, S., Ishibashi, M., Tsuyuki, T., Takahashi, T., Honda, T., and Nakanishi, T. (1984). *Chem. Pharm. Bull.* **32**, 4702.

Sakaki, T., Yoshimura, S., Ishibashi, M., Tsuyuki, T., Takahashi, T., Honda, T., and Nakanishi, T. (1985). *Bull. Chem. Soc. Jpn.* **58**, 2680.

Sakan, K., and Craven, B. M. (1983). *J. Am. Chem. Soc.* **105**, 3732.

Sasamori, H., Reddy, K. S., Kirkup, M. P., Shabanowitz, J., Lynn, D. G., Hecht, S. M., Woode, K. A., Bryan, R. F., Campbell, J., Lynn, W. S., Egert, E., and Sheldrick, G. M. (1983). *J. Chem. Soc., Perkin Trans.* **1**, 1333.

Sastry, K. V., and Rao, E. V. (1983). *Planta Med.* **47**, 227.

Saxton, W. M., Stemple, D. L., and McIntosh, J. R. (1984). *J. Cell. Biol.* **99**, 38a.

Schultz, G. A. (1973). *Chem. Rev.* **73**, 385.

Schun, Y., and Cordell, G. A. (1985). *J. Nat. Prod.* **48**, 684.

Segal, S. J. (1950). "Gossypol, A Potential Contraceptive for Men". Plenum Press, New York.

Segall, H. J., Wilson, D. W., Dallas, J. L., and Haddon, W. F. (1985). *Science* **229**, 472.

Seguin, E., Trinh, M. C., Koch, M., Andre, S., Farjaudon, N., Pareyre, C., Tempete, C., Robert-Gero, M., and Bourot, C. (1985). *Ann. Pharm. Fr.* **43**, 301.

Senilh, V., Gueritte, F., Guenard, D., Colin, M., and Potier, P. (1984). *C. R. Acad. Sci., Ser. 2*, **299**, 1039.

Shamma, M. (1972a). *In* "The Isoquinoline Alkaloids", pp. 426–457. Academic Press, New York.

Shamma, M. (1972b). *In* "The Isoquinoline Alkaloids", pp. 315–343. Academic Press, New York.

Shamma, M., and Georgiev, V. St. (1974). *J. Pharm. Sci.* **63**, 163.

Shamma, M., and Moniot, J. L. (1978a). *In* "Isoquinoline Alkaloids Research 1972–1977", pp. 355–363. Plenum Press, New York.

Shamma, M., and Moniot, J. L. (1978b). *In* "Isoquinoline Alkaloids Research 1972– 1977", pp. 271–292. Plenum Press, New York.

Shishido, K., Saitoh, T., Fukumoto, K., and Kametani, T. (1984). *J. Chem. Soc., Perkin Trans.* **1**, 2139.

Simanek, V. (1985). *In* "The Alkaloids, Vol. XXVI" (A. Brossi, ed.), pp. 185–240. Academic Press, New York.

Simonds, R., De Bruyn, A., De Taeye, L., Verzele, M., and De Pauw, C. (1984). *Planta Med.* **50**, 274.

Sinkule, T. A. (1984). *Pharmacotherapy* **4**, 61.

Smith, C. R. Jr., Mikolajczak, K. L., and Powell, R. G. (1980). *In* "Anticancer Agents Based on Natural Product Models" (J. M. Cassady, and J. D. Douros, eds.), p. 392. Academic Press, New York.

Smith, C. R., and Powell, R. G. (1984). *In* "Alkaloids, Chemical and Biological Perspectives, Vol. II" (S. W. Pelletier, ed.), pp. 149–203. John Wiley and Sons Inc., New York.

Spencer, G. F., and Flippen-Anderson, J. L. (1981). *Phytochemistry* **20**, 2757.

Steward, O., Goldschmidt, R. B., and Sutula, T. (1984). *Life Sci.* **35**, 43.

Suffness, M., and Cordell, G. A. (1985). *In* "The Alkaloids, Vol. XXV" (A. Brossi, ed.), pp. 1–369. Academic Press, New York.

Suffness, M., and Douros, J. D. (1980). *In* "Anticancer Agents Based on Natural Produc. Models" (J. M. Cassady and J. D. Douros, eds.), p. 465. Academic Press, New York.

Suryawashi, S. N., and Fuchs, P. L. (1986). *J. Org. Chem.* **51**, 902.

Szantay, Cs., Szabo, L., Honty, K., Nogradi, K., Dezseri, E., Dancsi, L., Lorincz, Cs., Szarvady, B., and Kovacs, L. (1982). Hungarian Patent, Teljes HU 24,149; *Chem. Abstr.* **99**, 64297d.

Taafrout, M., Rouessac, F., and Robin, J.-P. (1983). *Tetrahedron Lett.* **24**, 197.

Takano, S., Otaki, S., and Ogasawara, K. (1985). *J. Chem. Soc., Chem. Commun.* p. 485.

Taylor, W. I., and Farnsworth, N. R. (1975). "The *Catharanthus* Alkaloids". Marcel Dekker, New York.

Thies, P. W. (1985). *Pharm. Unserer Zeit* **14**, 149.

Thomas-Jacob, R., and Reddi, O. S. (1984). *Life Sci. Adv.* **3**, 92.

Toyama, Y. (1984). *Seitai No Kaguku* **35**, 530; *Chem. Abstr.* **102**, 197344n.

Tso, W. W. (1984). *Cancer Lett.* **24**, 257.

Tsumura Jutendo Co. (1983). Jpn. Kokai Tokkyo Koho Jp 58 41,874; *Chem. Abstr.* **99**, 139653.

Ueda, M., Doteuchi, M., Nakamura, M., Kawakami, M., Hirose, F., and Miyata, K. (1983). *Oyo Yakuri* **25**, 143; *Chem. Abstr.* **98**, 172607t.

Vidari, G., Ferrino, S., and Grieco, P. A. (1984). *J. Am. Chem. Soc.* **106**, 3539.

Wall, M. E., and Wani, M. W. (1980). *In* "Anticancer Agents Based on Natural Product Models" (M. Cassady and J. D. Douros, eds.), p. 417. Academic Press, New York.

Wall, M. E., and Wani, M. (1984). *Adv. Chin. Med. Mater. Res., Int. Symp.* p. 391; *Chem. Abstr.* **104**, 141737y.

Wang, C.-L. J., and Ripka, W. C. (1983). *J. Org. Chem.* **48**, 2555.

Wang, Y., Li, Y., Pan, X., Li, S., and Huang, W. (1985). *Huaxue Xuebao* **43**, 161; *Chem. Abstr.* **103**, 71569y.

Weinreb, S. M., and Semmelhack, M. F. (1975). *Acc. Chem. Res.* **8**, 158.

Weng, Z.-Y., Wang, Z.-Y., and Yan, X.-M. (1982). *Yaoxue Xuebao* **17**, 744.

Whiting, D. A. (1985). *Nat. Prod. Rep.* **2**, 191.

Wiedenfeld, H., Kirtel, A., and Roeder, E. (1985). *Arch. Pharm. (Weinheim)* **318**, 294.

Wiegrebe, W., Kramer, W. J., and Shamma, M. (1984). *J. Nat. Prod.* **47**, 397.
Willingham, W., Considine, R. T., Chaney, S. G., Lee, K. M., and Hall, I. H. (1984). *Biochem. Pharmacol.* **33**, 330.
Winograd, B., Vermorken, J. B., van Maanen, J., and Pinedo, H. M. (1984). *Pharm. Weekbl.* **119**, 1277.
Winterfeldt, E. (1975). *Recent Dev. Chem. Nat. Carbon Compd.* **6**, 9.
Wrobel, J. T. (1985). *In* "The Alkaloids, Vol. 26" (A. Brossi, ed.), pp. 327–384. Academic Press, New York.
Wu, G., Fang, F., and Zuo, J. (1984). *Yaoxue Xuebao* **19**, 167; *Chem. Abstr.* **103**, 16415v.
Xu, B., and Yang, J. (1984). *Adv. Chin. Med. Mater. Res., Int. Symp.* p. 377; *Chem. Abstr.* **104**, 179527t.
Xu, B., and Yang, J. (1985). *Yiyao Gongye* **16**, 550; *Chem. Abstr.* **104**, 199304w.
Yakult Honsha Co. (1985). Jpn. Kokai Tokkyo Koho JP 60 19,790; *Chem. Abstr.* **103**, 88119z.
Yamaguchi, H., Arimoto, M., Tanoguchi, T., Ishida, T., and Inoue, M. (1982). *Chem. Pharm. Bull.* **30**, 3213.
Yamaguchi, H., Nakajima, S., Arimoto, M., Tanoguchi, M., Ishida, T., and Inoue, M. (1984). *Chem. Pharm. Bull.* **32**, 1754.
Yamamoto, K., Hirose, K., Eigyo, M., Utsumi, S., Shiomi, T., Shintaku, H., Koshida, H., Takahara, Y., Kattau, R. W., *et al.* (1983). *Oyo Yakun* **25**, 133; *Chem. Abstr.* **98**, 172606s.
Yoshimura, S., Narita, H., Hayashi, K., and Mitsuhashi. H. (1983). *Chem. Pharm. Bull.* **31**, 3971.
Yoshimura, S., Sakaki, T., Ishibashi, M., Tsuyuki, T., Takahashi, T., and Honda, T. (1985). *Bull. Chem. Soc. Jpn.* **58**, 2673.
Yoshimura, S., Sakaki, T., Ishibashi, M., Tsuyuki, T., Takahashi, T., Matsushita, K., and Honda, T. (1984). *Chem. Pharm. Bull.* **32**, 4698.
Zalkov, L. H., Glinski, J. A., Gelbaum, L. T., Fleischmann, T. J., McGowan, L. S., and Gordon, M. M. (1985). *J. Med. Chem.* **28**, 687.
Zhang, J., Xu, R., Li, Y., and Chen, Z. (1984). *Huaxue Xuebao* **42**, 684; *Chem. Abstr.* **101**, 147837r.
Ziegler, F. E., Klein, S. I., Pati, U. K., and Wang, T.-F. (1985). *J. Am. Chem. Soc.* **107**, 2730.

Index